555 Questions in Veterinary and Tropical Parasitology

555 Questions in Veterinary and Tropical Parasitology

Hany M. Elsheikha, *BVSc, MSc, PhD, FRSPH, PGCHE, FHEA, DipEVPC*

Faculty of Medicine and Health Sciences, University of Nottingham, UK

Xing-Quan Zhu, *BVSc, MVSc, PhD*

State Key Laboratory of Veterinary Etiological Biology, Lanzhou Veterinary Research Institute, Chinese Academy of Agricultural Sciences, The People's Republic of China

CABI is a trading name of CAB International

CABI
Nosworthy Way
Wallingford
Oxfordshire OX10 8DE
UK

Tel: +44 (0)1491 832111
Fax: +44 (0)1491 833508
E-mail: info@cabi.org
Website: www.cabi.org

CABI
745 Atlantic Avenue
8th Floor
Boston, MA 02111
USA

Tel: +1 (617)682-9015
E-mail: cabi-nao@cabi.org

A catalogue record for this book is available from the British Library, London, UK.

ISBN-13: 978 1 78924 234 8 (paperback)
         978 1 78924 235 5 (ePDF)
         978 1 78924 236 2 (ePub)

Commissioning Editor: Alexandra Lainsbury
Editorial Assistant: Emma McCann
Production Editor: Shankari Wilford

Typeset by SPi, Pondicherry, India
Printed and bound in the UK by Severn, Gloucester

Contents

About the Authors

Hany M. Elsheikha, BVSc, MSc, PhD, FRSPH, PGCHE, FHEA, DipEVPC

Hany is an Associate Professor, University of Nottingham, UK. He earned his PhD in Molecular and Evolutionary Parasitology from Michigan State University. In 2005, he was awarded the National Center for Infectious Diseases (NCID), Centers for Disease Control and Prevention (CDC) Postdoctoral Fellowship. Dr Elsheikha has contributed more than 250 research articles, primarily related to parasite biology and control. He has served on the editorial board of numerous peer-reviewed journals and has lectured at universities and conferences throughout the world. He is the author or editor of four textbooks. Also, he is a diplomate of the European Veterinary Parasitology College, a member of the European Scientific Counsel of Companion Animal Parasites UK & Ireland, a Fellow of the Royal Society of Public Health, and a Fellow of the Higher Education Academy. Since 2007 he has been at the University of Nottingham, where he established the veterinary parasitology curriculum from its inception.

Xing-Quan Zhu, BVSc, MVSc, PhD

Xing-Quan is Professor and Head of Department of Parasitology, State Key Laboratory of Veterinary Etiological Biology, Lanzhou Veterinary Research Institute, Chinese Academy of Agricultural Sciences, The People's Republic of China. He obtained his PhD and acquired postdoctoral training in Molecular Parasitology from the University of Melbourne, Australia. Between 2002 and 2010, Prof. Zhu taught veterinary parasitology in South China Agricultural University. He serves as a Subject Editor for Parasites & Vectors and a Section Editor for Parasitology Research. He also serves on the editorial boards of Trends in Parasitology, Veterinary Parasitology, Experimental Parasitology and Journal of Helminthology. He has published more than 300 articles in well-regarded international journals.

Preface

Parasitology is a large and important branch of life science, and as such covers a large amount of the knowledge tested in the undergraduate and specialty board examinations. In recent years, this field has undergone a tremendous expansion of knowledge, with significant advances in the diagnosis and control of parasitic diseases. It is therefore important to provide a convenient and current source of information for those interested in learning, revising and assessing their knowledge in parasitology.

This was the motivation to produce the first edition of the *555 Questions in Veterinary and Tropical Parasitology*. Our aim was to provide essential information in a concise and intellectually stimulating manner in order to engage the readers with the content. This book includes 555 questions, representing some of the most important parasites parasitologists have ever known as well as those parasites that are neglected. The questions cover a wide range of topics about parasite biology, epidemiology, diagnostics and management, and encompass both basic parasitology concepts and skills, through to advanced topics such as evidence-based parasitology. The answers are provided for all questions at the end of each section, with additional comments or explanation, when necessary, to assist with understanding and learning about the topic and not just checking if the answer is right or wrong. We hope that you will find the questions useful in your revision and wish you good luck in your study and examination.

Parasitology will always be a fascinating subject; the readers will find that there is much more to learn about parasitic diseases than what they already know. It is hoped that this book serves all parasitology trainees as a source of practice questions and a rapid review of the subject.

Hany M. Elsheikha and Xing-Quan Zhu

Acknowledgements

The authors wish to acknowledge the support of many colleagues and students who provided constructive feedback during the production of this book.

Notes for Users

This book includes five sections, where each section contains a certain type of questions. There are 555 questions categorized into: 250 multiple choice questions, 100 matching questions, 100 fill-in-the-blank questions, 50 true/false questions, and 55 image-based questions. While some of the questions test knowledge, we have also tried to assess understanding and decision-making skills.

Multiple Choice Questions (MCQs)

The use of MCQs is common in formative and summative examination. MCQs are composed of one question (stem) with multiple possible answers, including the correct answer and several incorrect answers (distractors). One correct answer must be selected from possible responses to the stem. Some of the MCQs are ordering questions, where readers select the correct order of a series of items. For example, some questions include a list of events of the parasite's life cycle and the readers choose the answer that places these events in the correct chronological order.

Matching Questions

With extended matching questions (EMQs), there is a theme for each question. This can be about parasite agents, diseases, diagnostic tests or treatment options. Also, there will be a list of ten possible answers, called 'Options', marked A–J. In the cross matching questions readers are provided with two lists, denoted as column A and column B. They will be asked to pair items from the two columns that have the most relevant association.

Fill-in-the-blank Questions (FIBs)

The FIB questions consist of a phrase or a sentence with one or more blank spaces where the reader provides the missing word(s). FIB-type questions differ from the other question types in that they are more objective and demand recall skills, making them a useful tool to assess acquisition and retention of knowledge.

True or False Questions

In this format of questions, readers choose true or false in response to a statement question.

Image-based Questions

With this question type the readers are presented with an image and then asked to identify particular structures as the answer. Also, we have included some illustrations to assess key morphological features of the parasites or key stages in their life cycles.

1 Multiple Choice Questions

Select the single correct answer(s) from the five options provided.

1. A 4-month-old kitten presents with cough, dyspnoea and nasal discharge. What is the most likely diagnosis?

   A. *Aelurostrongylus abstrusus*

   B. *Toxocara cati*

   C. *Taenia taenieaformis*

   D. *Ancylostoma tubaeforme*

   E. *Dipylidium caninum*

2. Infection with *Balantidium coli* occurs more often in communities that live in proximity to?

   A. Cows

   B. Horses

   C. Pigs

   D. Sheep

   E. Dogs

3. In regard to toxoplasmosis, which test can be used to discriminate recent from past infection?

   A. Serological detection of IgA

   B. Serological detection of IgD

C. Serological detection of IgG avidity

D. Serological detection of IgM

E. Serological detection of IgE

4. Post-mortem examination of a Roach fish with deformed abdomen and displaced fins revealed the following structures. What is the most likely diagnosis?

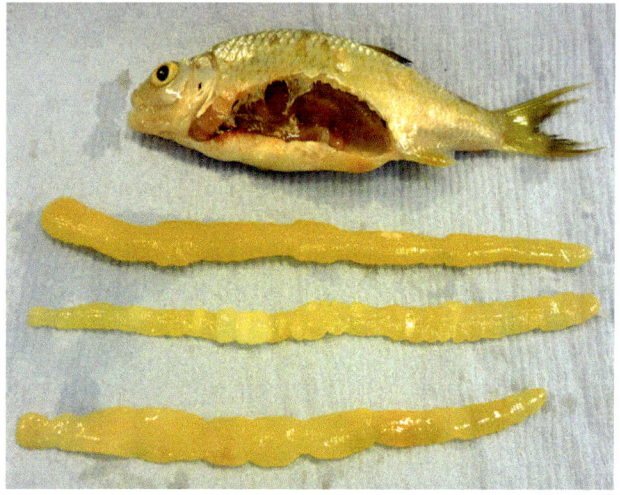

Fig. 1.1. Image courtesy of Emma Drinkall, University of Nottingham.

A. *Ligula intestinalis*

B. *Diphyllobothrium latum*

C. *Anisakis simplex*

D. *Anguillicola crassus*

E. *Contracaecum bancrofti*

5. Which one of the following cytokines reduces the inflammatory Th1 response during acute *Toxoplasma gondii* infection?

A. Tumor necrosis factor alpha (TNF-α)

B. Interferon gamma (IFN-γ)

C. Interleukin 1 (IL-1)

D. Interleukin 6 (IL-6)

E. Interleukin 10 (IL-10)

6. Which one of the following tapeworm species forms hydatid cysts in humans and sheep?

A. *Echinococcus granulosus*

B. *Echinococcus multilocularis*

C. *Taenia saginata*

D. *Taenia hydatigena*

E. *Taenia ovis*

7. Which one of the following systems is most impaired in patients with *Wuchereria bancrofti* infection?

A. Lymphatic system

B. Nervous system

C. Ocular system

D. Endocrine system

E. Musculoskeletal system

8. Dogs may contract mites via:

A. Eating faeces infected with mite eggs

B. Direct contact with an infested animal

C. Fomites (inanimate objects, such as a brush or comb)

D. Airborne infection

E. Both B and C

9. Barber's pole worm is the common name of _____.

A. *Chabertia ovina*

B. *Teladorsagia circumcincta*

C. *Haemonchus contortus*

D. *Nematodirus battus*

E. *Nematodirus spathiger*

10. Which one of the following is an abortifacient parasite in cattle?

A. *Haemonchus contortus*

B. *Toxocara vitulorum*

C. *Neospora caninum*

D. *Cryptosporidium bovis*

E. *Fasciola hepatica*

11. Which of the following is the infective stage of *Fasciola hepatica*?

A. Cercariae

B. Metacercariae

C. Rediae

D. Miracidia

E. Sporocysts

12. Infection with _____ causes a venereal disease in cattle.

A. *Tritrichomonas foetus*

B. *Toxoplasma gondii*

C. *Cooperia punctata*

D. *Babesia bovis*

E. *Schistosome mansoni*

13. What is the correct sequence of life cycle stages of *Plasmodium* species inside humans?

A. Sporozoite, ring-stage, trophozoite, merozoite, schizont, gametocyte

B. Sporozoite, merozoite, ring-stage, schizont, gametocyte, trophozoite

C. Sporozoite, gametocytes, merozoite, ring-stage, trophozoite, schizont

D. Sporozoite, merozoite, ring-stage, gametocyte, trophozoite, schizont

E. Sporozoite, merozoite, ring-stage, trophozoite, schizont, gametocyte

14. _____ has been incriminated as a possible risk factor for the development of 'nodding syndrome'.

A. *Ancylostoma duodenale*

B. *Onchocerca volvulus*

C. *Naegleria fowleri*

D. *Plasmodium vivax*

E. *Toxoplasma gondii*

15. Dourine is mainly reported in which of the following European countries?

A. France

B. Spain

C. Italy

D. United Kingdom

E. Greece

16. In which country was the canine lungworm *Angiostrongylus vasorum* first detected?

A. France

B. England

C. Ireland

D. Spain

E. Turkey

17. Autoinfection is a notable feature in the life cycle of _____.

A. *Cryptosporidium*

B. *Giardia lamblia*

C. *Balantidium coli*

D. *Taenia saginata*

E. *Enterobius vermicularis*

18. Who were the scientist(s) awarded the 2015 Nobel Prize in Physiology or Medicine?

A. Youyou Tu

B. William C. Campbell

C. Satoshi Ōmura

D. A and B

E. A, B and C

19. Found in dog faeces on a faecal float. What is it?

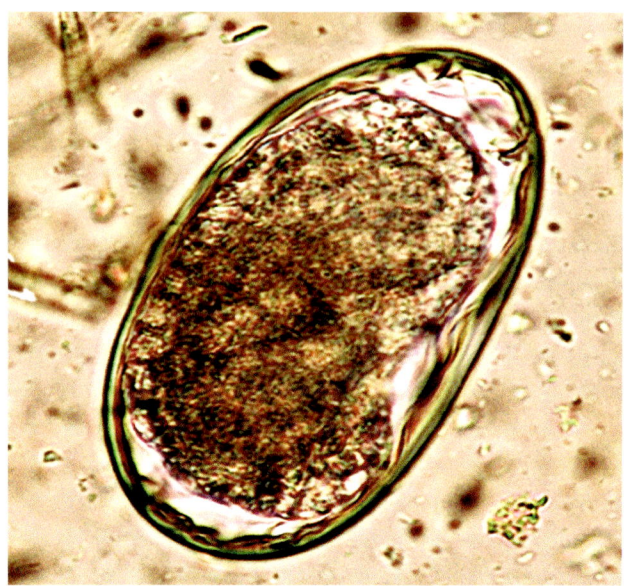

Fig. 1.2.

  A. *Ancylostoma caninum*

  B. *Dipylidium caninum*

  C. *Taenia solium*

  D. *Taenia saginata*

  E. *Toxocara canis*

20. Black flies act as vectors of _____.

  A. *Onchocerca volvulus*

  B. *Dirofilaria tenuis*

  C. *Dipetalonema evansi*

  D. *Toxocara canis*

  E. *Strongyloides stercoralis*

21. *Biomphalaria* snails are vectors of _____.

  A. *Schistosoma mansoni*

  B. *Ornithobilharzia* species

C. *Macrobilharzia* species

D. *Schistosomatium* species

E. *Schistosoma haematobium*

22. Which one of the following lists of phases of *Eimeria* life cycle is in the right order?

A. Sporozoite, oocyst, gamete, zygote, merozoite

B. Zygote, gametes, merozoite, oocyst, sporozoites

C. Oocyst, sporozoite, merozoite, gamete, zygote

D. Merozoite, sporozoite, oocyst, gamete, zygote

E. Oocyst, zygote, merozoite, sporozoite, gametes

23. _____ is the larval metacestode of *Taenia hydatigena*.

A. *Cysticercus ovis*

B. *Cysticercus tenuicollis*

C. *Cysticercus pisiformis*

D. *Cysticercus bovis*

E. *Cysticercus cellulosae*

24. *Strobilocercus fasciolaris* is found in _____.

A. Cats

B. Dogs

C. Ferrets

D. Rats

E. Rabbits

25. Parthenogenesis is a type of reproduction that occurs in _____.

A. *Styrongyloides* spp.

B. Ascarids spp.

C. *Ancylostoma* spp.

D. *Spirometra* spp.

E. *Trichostrongylus* spp.

26. _____ lives independent of a host, but may occasionally be parasitic under certain conditions.

    A. Erratic parasite

    B. Obligate parasite

    C. Facultative parasite

    D. Accidental parasite

    E. Ectoparasite

27. Oribatid mites serve as intermediate hosts for _____.

    A. _Anoplocephala perfoliata_

    B. _Oxyuris equi_

    C. _Strongylus vulgaris_

    D. _Strongylus equinus_

    E. _Strongylus edentatus_

28. _____ is the stage when a parasite can invade a host body and colonize it.

    A. Vegetative

    B. Latent

    C. Trophozoite

    D. Juvenile

    E. Infective

29. Which of these drugs can clear the latent liver stage (hypono-zoites) of _Plasmodium vivax_?

    A. Quinine

    B. Chloroquine

    C. Artemisinin-based combination therapies (ACTs)

    D. Tafenoquine (TQ)

    E. Doxycycline

30. The 'Rat-tailed' appearance in horses is caused by _____.

    A. _Parascaris equorum_

    B. _Strongylus vulgaris_

C. *Oxyuris equi*

D. *Strongylus equinus*

E. *Strongylus edentatus*

31. *Toxoplasma gondii* infection is acquired mainly through ingestion of:

A. Tissue cysts in undercooked pork

B. Vegetables contaminated with oocysts

C. Water containing oocysts

D. Fruits contaminated with oocysts

E. All of the above

32. Lateral uterine branches in gravid proglottids are counted for diagnosis of infection caused by _____.

A. *Diphyllobothrium latum*

B. *Dipylidium caninum*

C. *Taenia saginata*

D. *Hymenolepis nana*

E. *Echinococcus granulosus*

33. Which one of the following is the infective stage of the nematode *Nematodirus*?

A. First-stage larva (L1)

B. Third-stage larva (L3)

C. Egg containing L1

D. Egg containing L2

E. Egg containing L3

34. _____ occurs when larvae accidentally infect non-natural hosts wherein they cannot mature into adults but instead wander randomly and cause extensive pathologies.

A. Hypobiosis

B. Phenotypic switching

C. Larva migrans

D. Metabolic switching

E. Catabolism

35. Hexagonal basis capituli, short mouthparts (hypostome and palpi), presence of festoons along the posterior margin, and eyes on the side of the scutum, are diagnostic features of which one of the following ticks?

A. *Ixodes scapularis*

B. *Dermacentor variabilis*

C. *Rhipicephalus sanguineus*

D. *Amblyomma americanum*

E. *Otobius megnini*

36. What are the main innate effector cells responsible for killing *Leishmania*?

A. Macrophages

B. Neutrophils

C. T lymphocytes

D. B lymphocytes

E. Natural killer cells

37. _____ is commonly known as 'gapeworm' and can cause a condition in poultry called 'gapes', where infected birds show breathing difficulties and stretching of the neck, gasping for breath.

A. *Ascaridia galli*

B. *Capillaria annulata*

C. *Syngamus trachea*

D. *Dispharynx nasuta*

E. *Heterakis gallinarum*

38. Cysticerci, which develop in fleas, are the intermediate stage of which one of the following tapeworms?

A. *Taenia serialis*

B. *Dipylidium caninum*

C. *Taenia multiceps*

D. *Taenia saginata*

E. *Echinococcus granulosus*

39. **This organism was detected on honey bees. What is it?**

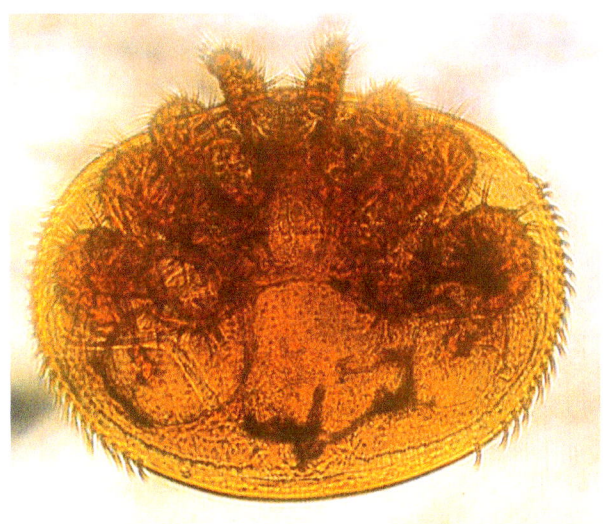

A. Varroa mite

B. Harvest mite

C. Dust mite

D. Plant-feeding mite

E. Trombiculid mite

40. **French lungworm disease in dogs is best diagnosed by** _____.

A. Direct faecal smear

B. Baermann technique

C. McMaster technique

D. Faecal culture

E. The modified Knott's method

41. _____ is most likely to interact with and increase the risk of ivermectin neurotoxicity in dogs?

    A. Albendazole

    B. Spinosad

    C. Metronidazole

    D. Selamectin

    E. Azithromycin

42. _____ is the infective stage of *Schistosoma* spp.

    A. Cercariae

    B. Miracidia

    C. Rediae

    D. Encysted metacercariae

    E. Embryonated eggs

43. Which organ is mainly damaged in sheep infected with ovine haemonchosis or teladorsagiosis?

    A. Small intestine

    B. Abomasum

    C. Large intestine

    D. Kidney

    E. Liver

44. Regarding fleas, which one of the following statements is correct?

    A. Removing pets from a flea-infested house for a while can help solve flea infestation

    B. Leave house vacant for months to starve fleas

    C. Fleas cannot survive the winter temperatures

    D. Pupae hatch when they sense vibrations

    E. Immature fleas can lie dormant for up to a month

45. One application of _____ can prevent the development of immature *Angiostrongylus vasorum* stages for 4 weeks.

    A. Milbemax®

    B. Panacur®

    C. Advocate®

    D. NexGard Spectra®

    E. Trifexis®

46. Which list of stages of *Fasciola hepatica* life cycle is in the right order?

    A. Egg, redia miracidium, sporocysts, cercaria, adult fluke, metacercariae

    B. Egg, cercaria, sporocysts, redia, metacercariae, adult fluke, miracidium

    C. Egg, miracidium, sporocysts, redia, cercaria, metacercariae, adult fluke

    D. Egg, sporocysts, redia, cercaria, metacercariae, adult fluke, miracidium

    E. Egg, metacercariae, sporocysts, miracidium, redia, cercaria, adult fluke

47. Which one of the following anthelmintics is incorrectly matched with its mechanism of action?

    A. Piperazine: cholinergic receptor antagonist/GABA receptor agonist

    B. Fenbendazole: binds to tubulin

    C. Levamisole: cholinergic receptor agonist

    D. Pyrantel: GABA receptor antagonist

    E. Organophosphates: indirect-acting cholinergic agonists

48. Although not a lungworm, _____ can induce transient respiratory manifestations in infected pigs due to its extra-intestinal migratory behaviour.

    A. *Ascaris suum*

    B. *Hyostrongylus rubidus*

    C. *Trichuris suis*

    D. *Trichinella spiralis*

    E. *Oesophagostomum dentatum*

49. Benznidazole is more effective at curing 'Chagas' disease during which of the following phases of the disease?

A. Acute phase

B. Chronic phase

C. Both acute and chronic phase

D. Neither acute nor chronic phase

E. Subacute phase

50. The control of _____ can be complicated by hepatic stage, which persists dormant in the liver even after the blood stages are cleared.

A. *Plasmodium vivax*

B. *Plasmodium falciparum*

C. *Plasmodium knowlesi*

D. *Plasmodium malariae*

E. *Plasmodium ovale*

51. *Hyostrongylus rubidus* worms are found in the stomach of _____.

A. Cattle

B. Sheep

C. Pigs

D. Camels

E. Llamas

52. The condition 'cercarial dermatitis' is caused by cercariae of which one of the following parasites?

A. *Schistosoma spindale*

B. *Schistosoma mansoni*

C. *Schistosoma haematobium*

D. *Schistosoma japonicum*

E. *Schistosoma mekongi*

53. This was detected in a urine microscope sample from an apparently healthy cat. What is it?

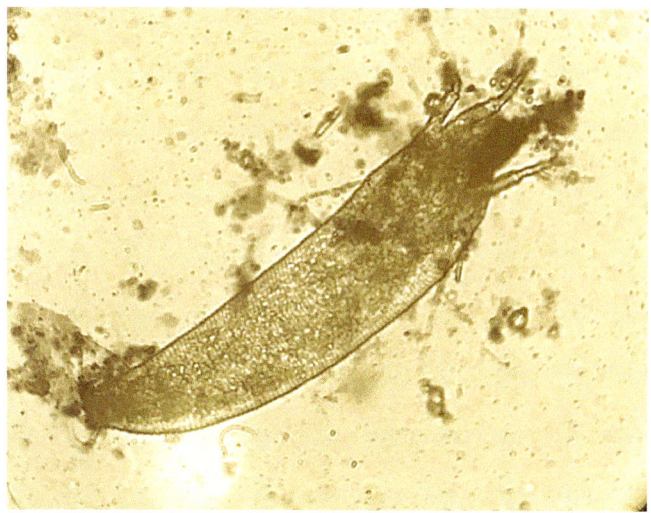

Fig. 1.4.

    A. *Phytoptus* species

    B. *Demodex gatoi*

    C. *Demodex cati*

    D. *Notoedres cati*

    E. *Otodectes cynotis*

54. Histological examination of the liver of a 4-year-old Thoroughbred shows granulomatous lesions (~1.5 cm in diameter), which contain nematode eggs surrounded by a fibrous capsule. The eggs have a barrel-like shape with an opercular plug at both ends and a double-layered shell, and measure ~50 × 20 µm. Which one of the following parasites is most likely involved in this case?

    A. *Capillaria hepatica*

    B. *Fasciola hepatica*

    C. *Echinococcus equinus*

    D. *Oxyuris equi*

    E. *Parascaris equorum*

55. Cobblestone lesions are found in the gastric mucosa of sheep infected with _____.

    A. *Trichostrongylus axei*

    B. *Teladorsagia circumcincta*

    C. *Haemonchus contortus*

    D. *Nematodirus battus*

    E. *Oesophagostomum columbianum*

56. The life cycle stages of mites progress in which one of the following sequences?

    A. Egg, larva, protonymph, deutonymph, tritonymph, adult

    B. Larva, tritonymph, deutonymph, protonymph, egg, adult

    C. Larva, protonymph, tritonymph, deutonymph, adult, egg

    D. Egg, adult, deutonymph, tritonymph, protonymph larva

    E. Tritonymph, deutonymph, protonymph egg, larva, adult

57. *Strongylus* spp. in equines are commonly known as _____.

    A. Red worms

    B. Brown worms

    C. Black worms

    D. Yellow worms

    E. Blue worms

58. Recycling of infective stage within the same host, without exiting the body, is called _____.

    A. Autoinfection

    B. Degradation

    C. Duplication

    D. Hypobiosis

    E. Autophagy

59. Black scour is associated with which one of the following parasites?

    A. *Trichostrongylus axei*

    B. *Trichuris ovis*

C. *Fasciola hepatica*

D. *Haemonchus contortus*

E. *Taenia solium*

60. All of the following are anti-*Angiostrongylus vasorum* drugs, except?

A. Prinovox®

B. Milbemax®

C. Advocate®

D. Milbactor®

E. Drontal®

61. Nodular worm is the common name of which one of the following parasite species?

A. *Trichuris ovis*

B. *Oesophagostomum dentatum*

C. *Haemonchus contortus*

D. *Taenia solium*

E. *Nematodirus* species

62. Which of the following is the scientific name of the equine pinworm?

A. *Habronema majus*

B. *Oxyuris equi*

C. *Habronema muscae*

D. *Draschia megastoma*

E. *Parascaris equorum*

63. Ferrets can be protected from toxoplasmosis by:

A. Avoiding feeding ferrets uncooked meat

B. Avoiding feeding ferrets meat by-products

C. Not allowing ferrets to eat prey (e.g. mice, rats, rabbits)

D. Keeping cat faeces from contaminating ferret foods, litter boxes, and housing areas

E. All of the above

64. Female *Toxocara canis* worms may produce up to _____.

    A. 500 eggs per day

    B. 5,000 eggs per day

    C. 50,000 eggs per day

    D. 100,000 eggs per day

    E. 200,000 eggs per day

65. Which one of the following drugs is effective against the multidrug-resistant *Plasmodium falciparum* and *P. vivax*?

    A. Doxycycline

    B. Gentamicin

    C. Clarithromycin

    D. Penicillin

    E. Ciprofloxacin

66. Which of the following measures can be used to prevent lyme disease?

    A. Avoiding tick-endemic areas

    B. Covering bare skin with white clothes

    C. Using tick repellents on clothing

    D. Removing ticks from skin as soon as possible

    E. All of the above

67. Blowfly strike caused by *Lucillia cuprina* and *L. sericata* is a major problem for sheep producers in?

    A. Nigeria, Ghana, and Niger

    B. Great Britain, Australia, and New Zealand

    C. China, Mongolia, and North Korea

    D. Colombia, Venezuela, and Cuba

    E. USA and Canada

68. A 25-year-old presented complaining of eye redness, photophobia, tearing and severe pain. A fluorescence microscopic examination of a corneal scraping stained with Calcofluor White

dye revealed rounded cyst-like structures, 10–20 μm in diameter. What is the most likely diagnosis?

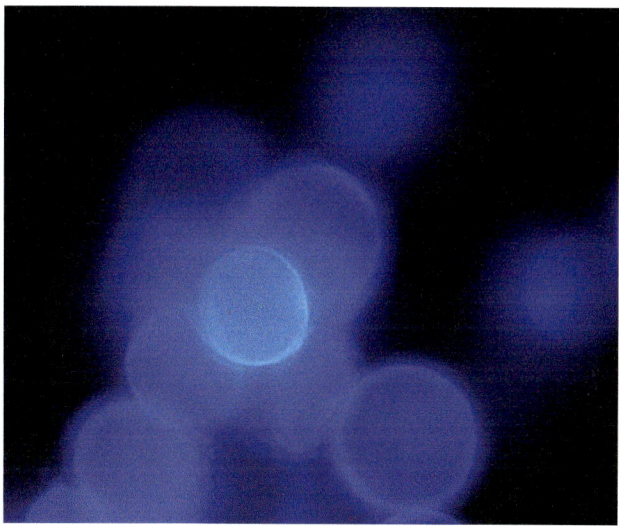

Fig. 1.5.

  A. Ocular toxoplasmosis

  B. Ocular larval migrans

  C. *Acanthamoeba* keratitis

  D. Ocular cysticercosis

  E. Ocular filariasis

69. Which one of the following was the first drug used in trials in human onchocerciasis to deplete *Wolbachia*?

  A. Doxycycline

  B. Gentamicin

  C. Clarithromycin

  D. Penicillin

  E. Ciprofloxacin

70. Which one of the following *Plasmodium* species is responsible for the most deadly form of human malaria?

  A. *Plasmodium vivax*

  B. *Plasmodium ovale*

C. *Plasmodium malariae*

D. *Plasmodium falciparum*

E. *Plasmodium knowlesi*

71. All the following are correct features of fenbendazole except?

    A. Is a benzimidazole antiparasitic agent

    B. Disturbs nervous stimuli transmission to muscles in nematodes

    C. Active against cestodes and nematodes, and some trematodes

    D. The presence of food in the gut can increase its availability

    E. May cause pancytopenia

72. A number of small (10–20 µm × 5–15 µm), pear-shaped, bilaterally symmetrical trophozoites were observed in the faeces of a cat with a history of diarrhoea. The trophozoites have two nuclei, eight flagella and a large sucking disc on the ventral surface at the broader end. What are these trophozoites?

    A. *Giardia intestinalis*

    B. *Tritrichomonas foetus*

    C. *Isospora cati*

    D. *Entamoeba histolytica*

    E. *Cryptosporidium felis*

73. Which one of the following is the only animal species that can actively shed *Toxoplasma gondii* oocysts?

    A. Squirrel

    B. Dog

    C. Cat

    D. Wolf

    E. Red fox

74. Which one of the following treatments is indicated for the management of pulmonary hypertension in a dog with lungworms and pre-existing renal disease?

    A. Fluid therapy

    B. Blood transfusion

C. Angiotensin converting enzyme (ACE) inhibitor

D. Anthelmintic

E. Antibiotics

75. _____ can potentiate the ivermectin-induced suppression of microfilaria in onchocerciasis patients.

A. Amoxicillin

B. Doxycycline

C. Ciprofloxacin

D. Metronidazole

E. Clindamycin

76. When was *Angiostrongylus vasorum* first detected in the British Isles?

A. 1954

B. 1968

C. 1960

D. 1911

E. 1970

77. Which two parasites colonize host erythrocytes?

A. *Plasmodium* and *Babesia*

B. *Toxoplasma gondii* and *Plasmodium*

C. *Cryptosporidium* and *Babesia*

D. *Toxoplasma gondii* and *Cryptosporidium*

E. *Cryptosporidium* and *Plasmodium*

78. Adult whipworms can be found in the _____.

A. Small intestine

B. Stomach

C. Lungs

D. Heart

E. Large intestine

79. _____ undertakes hepato-pancreatic migration inside the horse.

    A. *Anoplocephala perfoliata*

    B. *Oxyuris equi*

    C. *Strongylus vulgaris*

    D. *Strongylus equinus*

    E. *Strongylus edentatus*

80. Coughing and respiratory distress are observed in first-season grazing calves. What is the likely causative agent?

    A. *Dictyocaulus viviparus*

    B. *Dictyocaulus filaria*

    C. *Dictyocaulus arnfieldi*

    D. *Protostrongylus rufescens*

    E. *Muellerius capillaris*

81. Which one of the following includes the cardinal presentation of caval syndrome in dogs?

    A. Tricuspid regurgitation, haemoglobinuria, dark brown to black urine

    B. Weight loss, diarrhoea, difficulty breathing

    C. Inactivity, bulging chest, rapid breathing

    D. Seizures, blindness, secondary pneumonia

    E. Vomiting, diarrhoea, lethargy, anorexia

82. _____ are referred to as cyathostomes.

    A. *Triodontophorus* sp.

    B. Small strongyles

    C. *Strongylus edentatus*

    D. *Strongylus vulgaris*

    E. *Strongylus equinus*

83. _____ are the *Plasmodium* stage that exit from the liver.

    A. Sporozoite

    B. Merozoites

C. Male microgamete

D. Female macrogamete

E. Hyponozoites

84. Which of the following parasites can increase the risk of developing colorectal cancer?

A. *Schistosoma japonicum*

B. *Trichutis trichiura*

C. *Ancylostoma tubaeforme*

D. *Ascaris lumbricoides*

E. *Heterophyes heterophyes*

85. Infected mosquitoes transmit which one of the following *Plasmodium* stages to humans?

A. Sporozoite

B. Merozoite

C. Male microgamete

D. Female macrogamete

E. Hyponozoite

86. What is the role of frog in the life cycle of *Angiostrongylus vasorum*?

A. Mechanical vector

B. Paratenic host

C. Intermediate host

D. Biological vector

E. Definitive host

87. A cough in young dogs can be caused by all of the following parasites except?

A. *Angiostrongylus vasorum*

B. *Ancylostoma caninum*

C. *Strongyloides* species

D. *Taenia hydatigena*

E. *Toxocara canis*

88. Which of these are considered a high-risk group for toxocariasis?

    A. Elderly people

    B. Pregnant women

    C. Immunocompromised patients with contact with dogs

    D. Children with geophagia

    E. All of the above

89. Which one of the following ticks is called 'brown tick'?

    A. *Rhipicephalus sanguineus*

    B. *Ixodes canisuga*

    C. *Dermacentor reticulatus*

    D. *Haemaphysalis punctata*

    E. *Ixodes hexagonus*

90. _____ is a contraindicated procedure in dogs suffering from high intracranial pressure.

    A. Cerebrospinal fluid analysis

    B. Magnetic resonance imaging

    C. Bronchoalveolar lavage

    D. Anterior chamber paracentesis

    E. Blood transfusion

91. Which one of the following is the correct sequence of the life cycle stages of *Plasmodium* sp. inside mosquitoes?

    A. Gametocytes, oocyst, zygote, ookinete, sporozoite

    B. Gametocytes, zygote, oocyst, sporozoite, ookinete

    C. Gametocytes, sporozoite, zygote, ookinete, oocyst

    D. Gametocytes, zygote, ookinete, oocyst, sporozoite

    E. Gametocytes, ookinete, sporozoite, oocyst, zygote

92. The best option for definitive diagnosis in an outbreak of acute fasciolosis is _____.

    A. Post mortem examination

    B. Faecal sedimentation test

   C. McMaster egg counting

   D. Serum antibody assay

   E. Copro-antigen detection

93. _____ **produces necrotic whitish lesions known as 'milk spots' on the liver capsule of pigs.**

   A. *Metastrongylus apri*

   B. *Ascaris suum*

   C. *Oesophagostomum dentatum*

   D. *Echinococcus granulosus*

   E. *Trichinella spiralis*

94. 'Slime ball' is a stage in the life cycle of which of the following parasites?

   A. *Dicrocoelium dendriticum*

   B. *Fasciola hepatica*

   C. *Schistosoma mansoni*

   D. *Schistosoma haematobium*

   E. *Paramphistomum cervi*

95. *Plasmodium falciparum* strain of malaria infects people mostly in _____.

   A. Africa

   B. Southwest Asia

   C. Central America

   D. Australia

   E. Southern Europe

96. Repeated treatments with oxibendazole failed to improve the condition of a Thoroughbred mare infected with *Strongylus vulgaris*, suggesting that the worms are tolerant to this drug. If you consider using another anthelmintic to treat the mare for this infection, which one of the following drugs should be avoided?

   A. Fenbendazole

   B. Ivermectin

C. Moxidectin

D. Pyrantel

E. Praziquantel

97. All of the following cestodes require an intermediate host to complete their development except?

A. *Echinococcus granulosus*

B. *Dipylidium caninum*

C. *Hymenolepis diminuta*

D. *Echinococcus multilocularis*

E. *Hymenolepis nana*

98. The life expectancy of adult botflies is _____.

A. A few weeks

B. A few days

C. A few months

D. Up to 1 year

E. Up to 1.5 years

99. Medium size (50–60 μm long and 30–35 μm wide), barrel-shaped, asymmetrical plugs, and rough shell with pitted surface, are all characteristic features of _____ eggs.

A. *Capillaria aerophila*

B. *Capillaria hepatica*

C. *Capillaria boehmi*

D. *Capillaria plica*

E. *Trichuris vulpis*

100. Which one of the following organs is most commonly involved in hydatid disease?

A. Liver

B. Heart

C. Kidney

D. Brain

E. Spleen

101. Which one of the following parasites is a risk factor for the development of lymphomas?

A. *Trichuris trichiura*

B. *Trichinella spiralis*

C. *Ascaris lumbricoides*

D. *Enterobius vermicularis*

E. *Strongyloides stercoralis*

102. *Coenurus cerebralis*, which occurs in the brain of sheep and cattle, is the intermediate stage of _____.

A. *Taenia serialis*

B. *Taenia pisiformis*

C. *Taenia multiceps*

D. *Taenia saginata*

E. *Taenia ovis*

103. Post mortem examination of a young pigeon revealed yellowish lesions in the pharynx, oesophagus and crop. Pear-shaped trophozoites with four flagella, each arising from a basal body located at the anterior broad end, were isolated. What is the likely diagnosis?

A. *Trichomonas gallinaea*

B. *Histomonas meleagridis*

C. *Cryptosporidium* species

D. *Eimeria* species

E. *Isospora* species

104. Ultrasound and MRI usually help in the diagnosis of infection caused by _____.

A. *Dipylidium caninum*

B. *Hymenolepis diminuta*

C. *Echinococcus granulosus*

D. *Hymenolepis nana*

E. *Toxocara canis*

105. Dogs and cats can be infected by a number of cardiorespiratory parasites. The commonest super-family to which most canine and feline animal lungworms belong is _____.

A. Metastrongyloidea

B. Ancylostomatoidea

C. Strongyloidea

D. Trichostrongyloidea

E. Heligosomoidea

106. Found in dog faeces on a faecal float (Fig. 6.1). What is it?

A. *Taenia pisiformis*

B. *Taenia saginata*

C. *Taenia solium*

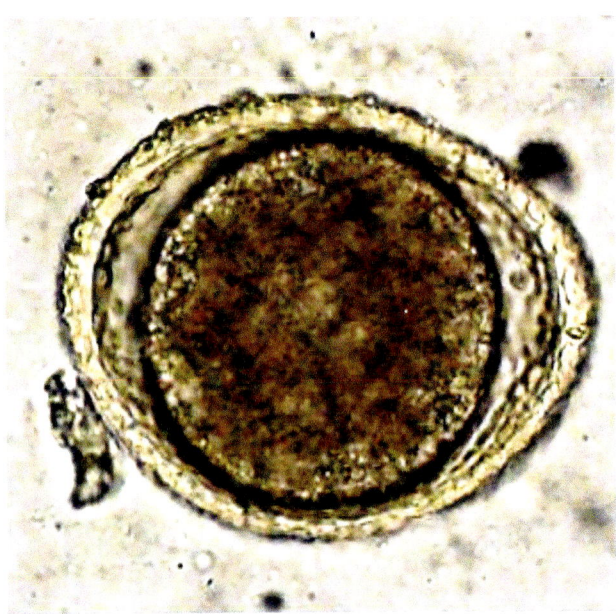

Fig. 1.6.

D. *Toxocara canis*

E. *Dipylidium caninum*

107. Tafenoquine (FDA-approved in 2018) is used for the treatment of individuals with _____.

A. *Plasmodium vivax*

B. *Plasmodium falciparum*

C. *Plasmodium ovale*

D. *Plasmodium knowlesi*

E. *Plasmodium malariae*

108. Which one of the following parasites produces ocular cysticercosis in humans?

A. *Dipylidium caninum*

B. *Taenia saginata*

C. *Taenia solium*

D. *Toxocara cati*

E. *Toxocara canis*

109. Which one of the following parasites increases the risk of developing hepatocellular carcinoma?

A. *Schistosoma mansoni*

B. *Balamuthia mandrillaris*

C. *Trichinella spiralis*

D. *Balantidium coli*

E. *Trichuris trichiura*

110. Which one of the following parasites can cause rectal prolapse in infected individuals?

A. *Ascaris lumbricoides*

B. *Balantidium coli*

C. *Trichuris trichiura*

D. *Strongyloides stercoralis*

E. *Necator americanus*

111. Inside the mosquito, *Plasmodium* gametes ultimately differentiate into _____.

    A. Sporozoite

    B. Merozoites

    C. Gametocytes

    D. Hyponozoite

    E. Ookinete

112. Tsetse flies act as vectors of _____.

    A. *Trypanosoma brucei*

    B. *Dipetalonema evansi*

    C. *Dirofilaria tenuis*

    D. *Babesia bigemina*

    E. *Dirofilaria immitis*

113. _____ causes intestinal schistosomiasis.

    A. *Schistosoma mansoni*

    B. *Schistosoma haematobium*

    C. *Austrobilharzia* species

    D. *Ornithobilharzia* species

    E. *Macrobilharzia* species

114. _____ is transmitted by the house fly.

    A. *Habronema muscae*

    B. *Dipetalonema evansi*

    C. *Dicrocoelium dendriticum*

    D. *Fasciola hepatica*

    E. *Fasciola gigantica*

115. Horse flies are the common name of members of which of the following insect families?

    A. Tabanidae

    B. Calliphoridae

C. Drosophilidae

D. Hippoboscidae

E. Oestridae

116. _____ can cause verminous (parasitic) thrombosis and arteritis in horses.

A. *Parascaris equorum*

B. *Strongylus vulgaris*

C. *Oxyuris equi*

D. *Strongylus equinus*

E. *Strongylus edentatus*

117. Which parasite can increase the risk of developing urinary bladder cancer?

A. *Schistosoma haematobium*

B. *Fasciola hepatica*

C. *Ascaris lumbricoides*

D. *Strongyloides stercoralis*

E. *Necator americanus*

118. _____ is the infective stage of *Strongylus edentates*.

A. First-stage larva (L1)

B. Third-stage larva (L3)

C. Egg containing L1

D. Egg containing L2

E. Egg containing L3

119. _____ is commonly used to diagnose infestation with the *Pneumonyssus caninum* mite.

A. Nasal swab

B. ELISA analysis

C. Endoscopy

D. Complete blood count

E. Immunofluorescence assay

120. _____ is transmitted by *Chrysops* fly.

    A. Loa loa

    B. *Dipetalonema evansi*

    C. *Dicrocoelium dendriticum*

    D. Guinea worm disease

    E. Rickettsiosis

121. Which one of the following parasites that affect sheep can also affect humans?

    A. *Trichostrongylus colubriformis*

    B. *Nematodirus battus*

    C. *Haemonchus contortus*

    D. *Trichuris ovis*

    E. *Teladrosagia circumcincta*

122. _____ is the scientific name of the neck threadworms.

    A. *Onchocerca reticulata*

    B. *Dicrocoelium dendriticum*

    C. *Dirofilaria tenuis*

    D. *Dirofilaria repens*

    E. *Setaria equina*

123. Mosquitoes act as vectors of _____.

    A. Malaria and Zika

    B. Tularaemia and relapsing fever

    C. African trypanosomiasis

    D. Typhus and relapsing fever

    E. Crimean-Congo haemorrhagic fever

124. With regards to *Cryptosporidium*, which of the following stages is responsible for autoinfection?

    A. Thick-walled oocyst

    B. Thin-walled oocyst

C. Microgamete

D. Macrogamont

E. Zygote

125. Arrested larval development occurs in which of the following parasites?

A. *Trichuris ovis*

B. *Haemonchus contortus*

C. *Fasciola hepatica*

D. *Toxascaris leonine*

E. *Taenia ovis*

126. Cyclops transmit which of the following parasites?

A. *Dracunculus medinensis*

B. *Dipetalonema evansi*

C. *Dirofilaria tenuis*

D. *Dirofilaria immitis*

E. *Dirofilaria repens*

127. _____ undertakes hepatic and peritoneal migration in horses.

A. *Anoplocephala perfoliata*

B. *Oxyuris equi*

C. *Strongylus vulgaris*

D. *Strongylus equinus*

E. *Strongylus edentates*

128. β-tubulin gene mutation often results in resistance against which of the following drugs?

A. Ivermectin

B. Benzimidazole

C. Levamisole

D. Moxidectin

E. Praziquantel

129. What is the first tissue to be affected following the bite by a malaria-infected mosquito?

    A. Spleen

    B. Kidney

    C. Liver

    D. Heart

    E. Eye

130. What factor in the epidemiology of haemonchosis contributes most to contamination of the pasture early in the year, providing a source of parasites for young lambs?

    A. Sharing of pasture by ewes and lambs

    B. Introduction of rams for breeding purposes

    C. Handling of lambs for docking and vaccination

    D. 'Spring rise', where eggs overwinter on pasture and ewes shed eggs because of periparturient relaxation of immunity

    E. Rotational grazing of sheep flocks

131. A dog nematode characterized by having its anterior extremity bent dorsally, what is it?

    A. *Toxocara canis*

    B. *Ancylostoma caninum*

    C. *Dirofilaria immitis*

    D. *Trichuris vulpis*

    E. *Toxascaris leonine*

132. Infection with _____ causes Guinea worm disease.

    A. *Dracunculus medinensis*

    B. *Dipetalonema evansi*

    C. *Dirofilaria tenuis*

    D. *Dictyocaulus filaria*

    E. *Mansonella perstans*

133. A sound worm management programme in a flock of sheep should include:

A. Rotation of pastures

B. Anthelminthic medication of pregnant ewes

C. Oral treatment of diarrhoeic animals

D. Separation of young and old animals

E. All of the above

134. Dicrocoeliasis is caused by _____.

A. The lancet liver fluke

B. The large American liver fluke

C. *Dipetalonema evansi*

D. *Dirofilaria tenuis*

E. *Dirofilaria immitis*

135. Management of acute and subacute fasciolosis can be best achieved by using _____.

A. Albendazole

B. Triclabendazole

C. Oxyclozanide

D. Closantel

E. Nitroxynil

136. A 3-year-old, neutered, male collie is presented for investigation of exercise intolerance. The dog had been in the USA two months ago. *Dirofilaria immitis* infection is suspected. How can you best manage this disease?

A. Control the mosquito vector

B. Keep dogs indoors, especially in endemic regions

C. Protect animals living in endemic/hyperendemic areas via screening and chemoprophylaxis

D. Avoid travelling to endemic regions

E. All of the above

137. _____ serves as a vector of *Histomonas meleagridis*.

    A. *Davainea proglottina*

    B. *Hetrakis gallinarum*

    C. *Ascaridia galli*

    D. *Amoebotaenia sphenoides*

    E. *Choanotaenia infundibulum*

138. Infection with _____ causes heartworm disease in dogs and cats.

    A. *Dirofilaria immitis*

    B. *Dipetalonema evansi*

    C. *Dirofilaria tenuis*

    D. *Acanthocheilonema reconditum*

    E. *Dirofilaria repens*

139. A 31-day-old Arabian Thoroughbred filly developed depression, unconsciousness, and unresponsiveness to stimulation, 12 hours after deworming. Which one of the following anthelmintics is most likely to be involved in this case?

    A. Fenbendazole

    B. Praziquantel

    C. Moxidectin

    D. Pyrantel

    E. Oxibendazole

140. A blood smear from a patient infected with _____ shows the presence of mulberry-shaped aggregates called 'morulae' within neutrophils.

    A. *Anaplasma*

    B. *Ehrlichia*

    C. *Hepatozoon*

    D. *Babesia canis*

    E. *Babesia nicroti*

141. _____ is used for parasite control on dairies because it requires no meat or milk withdrawal when used according to the label.

    A. Eprinomectin

    B. Ivermectin

    C. Moxidectin

    D. Doramectin

    E. Levamisole

142. _____ is the only *Taenia* sp. that affects cats.

    A. *Taenia taenieaformis*

    B. *Taenia solium*

    C. *Taenia hydatigena*

    D. *Taenia crassiceps*

    E. *Taenia serialis*

143. Terrestrial snails act as vectors of _____.

    A. *Dicrocoelium dendriticum*

    B. *Dirofilaria tenuis*

    C. *Dirofilaria immitis*

    D. *Dipylidium caninum*

    E. *Diphyllobothrium latum*

144. _____ is the infective stage of *Wuchereria bancrofti* (agent of lymphatic filariasis).

    A. Second-stage larva (L2)

    B. Filariform (third-stage) larva (L3)

    C. First-stage larva (L1)

    D. Onchosphere

    E. Cysticercoid

145. In utero infections do occur in _____.

    A. *Toxocara canis*

    B. *Toxocara cati*

C. *Toxascaris leonine*

D. *Ascaris suum*

E. *Strongylus vulgaris*

146. _____ **causes blackhead disease.**

A. *Histomonas meleagridis*

B. *Cryptosporidium baileyi*

C. *Eimeria tenella*

D. *Eimeria maxima*

E. *Trichomonas gallinae*

147. Anterior chamber paracentesis (ACP) refers to a procedure used to _____**.**

A. Isolate larvae from the eye

B. Examine blood vessels in the eye

C. Diagnose certain eye conditions

D. Determine the cause of dry eye

E. Treat cataract

148. *Taenia solium* **develops in the definitive host after ingestion of** _____**.**

A. Cysticercoid larva

B. *Cysticercus bovis*

C. *Cysticercus cellulosae*

D. *Taenia* egg

E. Hydatid cyst

149. _____ **is a common diagnostic technique used for diagnosis of** *Dictyocaulus viviparus* **infection in cattle.**

A. Baermann apparatus

B. X-ray

C. Ultrasound

D. MRI

E. Faecal culture

150. Which one of the following drugs is not a member of isoxazoline class of ectoparasiticides?

  A. Bravecto®

  B. NexGard®

  C. Simparica®

  D. Credelio®

  E. Comfortis®

151. All of the following are features of imidacloprid, except?

  A. A neonicotinoid insecticide

  B. Combined with flumethrin in a slow-release collar formulation

  C. Combined with moxidectin in a topical spot-on formulation

  D. Binds to post-synaptic nicotinic receptors resulting in paralysis and death

  E. Interacts with ligand-gated (GABA) chloride channels

152. High concentration of pepsinogen in blood plasma in sheep is indicative of infection with _____.

  A. *Ostertagia ostertagi*

  B. *Fasciola hepatica*

  C. *Muellerius capillaris*

  D. *Trichostrongylus colubriformis*

  E. *Nematodirus battus*

153. A 7-year-old Warmblood gelding presents in July with intense pruritic dermatitis. Which one of the following insect types is most likely involved in this condition?

  A. *Culicoides* (midges)

  B. *Ctenocephalides felis* (flea)

  C. *Culex* species (mosquito)

  D. *Gastrophilus* species (botfly)

  E. *Bovicola equi* (biting louse)

154. Which one of the following parasitic diseases has both a bacterial and a parasitic component?

    A. Equine dourine

    B. Ovine haemonchosis

    C. Feline giardiasis

    D. Canine heartworm

    E. Human cryptosporidiosis

155. _____ is a prominent post mortem finding in teladorsagiosis in sheep.

    A. Mucosal haemorrhage

    B. Intestinal preformation

    C. Moroccan leather appearance in abomasum

    D. Enteritis

    E. Enlarged mesenteric lymph nodes

156. _____ is the infective stage of raccoon roundworm, *Baylisascaris procyonis*.

    A. First-stage larva (L1)

    B. Second-stage larva (L2)

    C. Third-stage larva (L3)

    D. Egg containing L1

    E. Egg containing L2

157. _____ can serve as a paratenic host that contributes to the spread of *Aelurostrongylus abstrusus* infection in cats.

    A. Frogs

    B. Skunks

    C. Opossum

    D. Armadillo

    E. Hedgehog

158. What tick typically inhabits buildings and kennels?

A. *Rhipicephalus sanguineus*

B. *Dermacentor reticulatus*

C. *Ixodes ricinus*

D. *Amblyomma maculatum*

E. *Hyalomma marginatum*

159. _____ is the most pathogenic *Eimeria* species known to infect the domestic chicken.

A. *Eimeria tenella*

B. *Eimeria maxima*

C. *Eimeria acervulina*

D. *Eimeria praecox*

E. *Eimeria mitis*

160. What is the mechanism that underpins the development of 'bottle jaw' in sheep?

A. Hypoalbuminaemia

B. Hyperkalaemia

C. Reduced protein synthesis

D. Accumulation of unfolded proteins

E. Endoplasmic-reticulum (ER) stress response

161. A Thoroughbred filly presents with acute onset of a lateral deviation of the cervical vertebrae (scoliosis). Treatment with flunixin meglumine and dexamethasone failed to result in any improvement and euthanasia was elected due to poor prognosis. Pathological examination of the spinal cord section of C4–C5 revealed haemorrhage, inflammatory cell infiltration, and lesions consistent with aberrant parasite migration. What is the most likely diagnosis?

A. *Parascaris equorum*

B. *Oxyuris equi*

   C. *Parelaphostrongylus tenuis*

   D. *Strongylus vulgaris*

   E. *Triodontophorus serratus*

162. The infective stage of *Toxocara canis* is _____.

   A. Egg containing L1

   B. Egg containing L3

   C. Egg containing L2

   D. First-stage larva (L1)

   E. Third-stage larva (L3)

163. All of the following are true about insect growth regulator (IGR) except?

   A. Interfere with chitin synthesis

   B. Possess significant activity against adult fleas

   C. Mimic action of juvenile insect hormones

   D. Have low toxicity to vertebrates

   E. Used as part of flea control

164. _____ has been used for years as an immune modulator and anthelmintic.

   A. Piperazine

   B. Albendazole

   C. Levamisole

   D. Fenbendazole

   E. Moxidectin

165. A subset of replicating *Plasmodium* merozoites inside human erythrocytes differentiate into _____.

   A. Hyponozoite

   B. Sporozoite

   C. Merozoites

D. Gametocytes

E. Ookinete

166. **The infective stage of *Entamoeba histolytica* is** _____.

A. Mature trophozoites

B. Freshly formed cysts

C. Mature quadrinucleate cysts

D. Freshly voided trophozoites

E. Sporulated oocysts

167. **The presence of genal and pronotal combs, with the first and second spines of genal comb approximately of equal length, are characteristic features of which of the following flea species?**

A. *Pulex irritans*

B. *Ctenocephalides canis*

C. *Ctenocephalides felis*

D. *Ceratophyllus gallinae*

E. *Xenopsylla cheopis*

168. **Visual impairment and blindness are the main health complications of** _____.

A. Encephalitozoonosis

B. Onchocerciasis

C. Leishmaniosis

D. Dirofilariosis

E. Fasciolosis

169. **What test is commonly used to isolate larvae in a faecal sample?**

A. McMaster slide

B. Baermann apparatus

C. Angio Detect™

D. ELISA

E. Ultracentrifugation

170. Which of the following sentences best describes the eggs of nematodes in the family Ascarididae (Ascarids)?

    A. Round with thick shell

    B. Has a bi-polar plug

    C. Elongated with clear morula

    D. Has a single polar plug

    E. Elongated with no clear morula

171. The first definitive human case with *Toxoplasma gondii* was discovered in _____.

    A. 1923

    B. 1911

    C. 1963

    D. 1891

    E. 1911

172. Paragonimiasis is _____.

    A. Human lung fluke disease

    B. Human liver fluke disease

    C. Human intestinal fluke disease

    D. Human blood fluke disease

    E. None of the above

173. Effector proteins destined to be secreted from *Toxoplasma gondii* apicomplexan organelles are released in which of the following orders?

    A. Micronemes, rhoptries and dense granules

    B. Dense granules, rhoptries and micronemes

    C. Rhoptries, micronemes and dense granules

    D. Micronemes, dense granules and rhoptries

    E. Dense granules, micronemes and rhoptries

174. _____ is not a cyst-forming protozoan.

A. *Neospora caninum*

B. *Besnoitia besnoiti*

C. *Toxoplasma gondii*

D. *Eimeria tenella*

E. *Sarcocystis neurona*

175. Which one of the following is known as the rodent hook-worm?

A. *Hymenolepis diminuta*

B. *Mastophorus muris*

C. *Nippostrongylus brasiliensis*

D. *Trichuris muris*

E. *Syphacia* sp.

176. Trichomoniasis is a venereal disease that mainly affects _____ .

A. Men

B. Women

C. Both sexes

D. Neither men nor women

E. Transgenders

177. Which one of the following is most recommended for treatment of uro-genital trichomoniasis?

A. Metronidazole

B. Sulfadiazine

C. Pyrimethamine

D. Spiramycin

E. Oxytetracycline

178. Symptomatic stage of malaria is associated with the replication of _____.

    A. Sporozoite

    B. Merozoites

    C. Male microgamete

    D. Female macrogamete

    E. Hyponozoite

179. 'Pinkeye disease' in cattle is associated with which one of the following fly vectors?

    A. Stable fly

    B. Face fly

    C. House fly

    D. Horn fly

    E. Warble fly

180. _____ does not have teeth in its buccal capsule.

    A. *Strongylus vulgaris*

    B. *Strongylus equinus*

    C. *Strongylus edentates*

    D. *Small Strongyles*

    E. *Strongyloides westeri*

181. _____ caused a waterborne outbreak in Milwaukee, Wisconsin (USA), in 1993.

    A. *Cryptosporidium* species

    B. *Toxoplasma gondii*

    C. *Giardia duodenalis*

    D. *Legionella pneumophila*

    E. *Pseudomonas aeruginosa*

182. Which one of the following drugs is recommended for the treatment of cryptosporidiosis?

    A. Nitazoxanide

    B. Amoxicillin

C. Albendazole

D. Azithromycin

E. Piperazine

183. Which one of the following drugs is effective against the tissue cyst of *Toxoplasma gondii*?

A. Sulfadiazine

B. Spiramycin

C. Pyrimethamine

D. None

E. Sulfadimidine

184. In which year was *Cryptosporidium* originally described?

A. 1907

B. 1911

C. 1900

D. 1897

E. 1916

185. Whipworm is the common name of _____.

A. *Trichuris ovis*

B. *Oxyuris equi*

C. *Haemonchus contortus*

D. *Taenia solium*

E. *Nematodirus* species

186. _____ is transmitted by snail hosts of the genus *Bulinus*.

A. *Schistosoma haematobium*

B. *Bivitellobilharzia* species

C. *Ornithobilharzia* species

D. *Macrobilharzia* species

E. *Schistosomatium* species

187. The term 'assassin bugs' refers to insects of which insect family?

  A. Reduviidae

  B. Hippoboscidae

  C. Simuliidae

  D. Sarcophagidae

  E. Oestridae

188. Flesh flies is the common name for flies of which of the following families?

  A. Oestridae

  B. Tabanidae

  C. Hippoboscidae

  D. Simuliidae

  E. Sarcophagidae

189. Which of the following is used to count worm eggs in a faecal sample?

  A. McMaster slide

  B. Haemocytometer slide

  C. Flow cytometry

  D. Cell counter

  E. Baermann apparatus

190. Myalgia, difficulty in mastication, breathing and swallowing, and eosinophilia are key symptoms of infection with _____.

  A. *Trichinella spiralis*

  B. *Trichuris trichiura*

  C. *Ascaris lumbricoides*

  D. *Enterobius vermicularis*

  E. *Ancylostoma duodenale*

191. A 7-year-old Thoroughbred horse developed severe colic, abdominal oedema, pyrexia, and diarrhoea. A faecal floatation test was negative, but bright red worms were observed in the faeces. What is the likely mechanism underpinning this condition?

A. Emergence of dormant cyathostome larvae from the gut wall

B. Extraintestinal migration of *Strongylus edentates*

C. Extraintestinal migration of *Strongylus equinus*

D. Rupture of *Echinococcus granulosus* cyst

E. Rupture of *Cysticercus tenuicollis* (larval stage of *Taenia hydatigena*)

192. How do humans acquire toxoplasmosis?

A. Congenital/transplacental transmission

B. Ingestion of parasite cysts in undercooked meat

C. Ingestion of food/water containing cat-shed oocysts

D. B and C

E. A, B and C

193. The gynecophoric canal is found in _____.

A. *Schistosoma mansoni*

B. *Fasciola hepatica*

C. *Taenia asiatica*

D. *Diphyllobothrium latum*

E. *Sparganum mansoni*

194. Chyluria, which is characterized by the passage of intestinal lymph (chyle) in urine, can be caused by _____.

A. *Brugia malayi*

B. Lymphoma

C. Tuberculosis

D. Pregnancy

E. Retroperitoneal abscess

195. Rapid weight loss and diarrhoea containing mucus and blood flecks in a 5-week-old lamb is most suggestive to which of the following parasitic diseases?

A. Nematodirosis

B. Coccidosis

C. Haemonchosis

D. Teladosagiosis

E. Trichostrongylosis

196. Microscopic examination of a faecal float from a stray dog revealed the presence of thin-walled tetrasporocystic, dizoic, sporulated coccidian oocysts. What is your diagnosis?

A. *Eimeria canis*

B. *Neospora caninum*

C. *Sarcocystis canis*

D. *Toxoplasma gondii*

E. *Isospora canis*

197. Sperm-like structures found inside nodules in the gills of a catfish, what are these?

A. *Henneguya* sp.

B. *Myxospora* sp.

C. *Microspora* sp.

D. *Myxobolus* sp.

E. *Kudoa* sp.

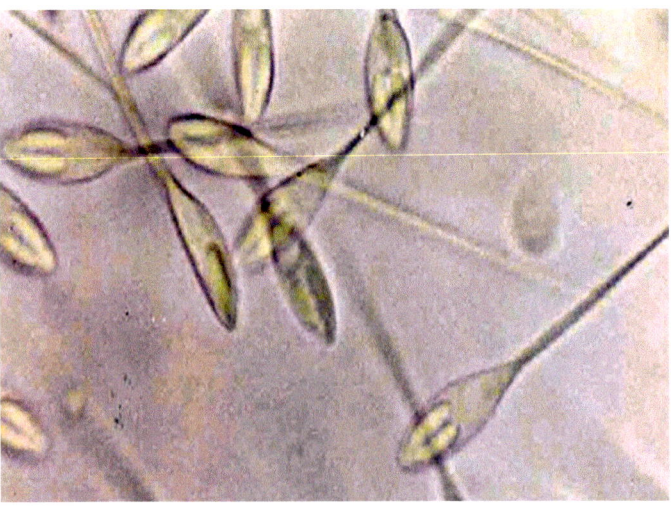

Fig. 1.7.

198. _____ has two rounded (ear-shaped) teeth in the buccal capsule.

    A. *Strongylus vulgaris*

    B. *Strongylus equinus*

    C. *Strongylus edentatus*

    D. *Small Strongyles*

    E. *Strongyloides westeri*

199. Patent *Strongyloides westeri* infections can develop in foals as young as _____.

    A. 1 week old

    B. 2 weeks old

    C. 3 weeks old

    D. 4 weeks old

    E. 5 weeks old

200. A link between exposure of Vietnam Veterans to parasite during their military service in Southeast Asia and development of cholangiocarcinoma five decades after primary infection has been suggested. What are the most likely incriminated parasites?

    A. *Schistosoma mansoni* and *S. japonicum*

    B. *Entamoeba histolytica* and *E. dispar*

    C. *Clonorchis sinensis* and *Opisthorchis viverrini*

    D. *Fasciola hepatica* and *F. gigantica*

    E. *Plasmodium falciparum* and *P. vivax*

201. Which one of the following abnormalities is caused by transdermal penetration of third stage larvae of *Strongyloides westeri* in foals?

    A. Frenzied behaviour

    B. Nose bleeding

    C. Oedema

    D. Hypersensitivity reaction

    E. Haemorrhage

202. Which one of the following parasites can reproduce in the environment without entering a host?

    A. *Strongyloides* spp.

    B. *Fasciola hepatica*

    C. *Ascaris suum*

    D. *Toxocara canis*

    E. *Trichuris ovis*

203. The transformation of trophozoite/vegetative stage into a cystic form is known as _____.

    A. Encystation

    B. Transformation

    C. Germination

    D. Trophozoitation

    E. Sporulation

204. Which of the following are the second most important vector of infectious diseases after mosquitoes?

    A. Bed bugs

    B. Ticks

    C. Lice

    D. Fleas

    E. Sandflies

205. Parthenogenetic females of _____ nematodes are capable of producing eggs on their own, without contribution from the male.

    A. *Strongyloides* spp.

    B. *Toxocara* spp.

    C. *Trichuris* spp.

    D. *Habronema* spp.

    E. *Ancylostoma* spp.

206. _____ is the causative agent of non-viral sexually transmitted disease.

   A. _Trichomonas vaginalis_

   B. _Toxoplasma gondii_

   C. _Leishmania infantum_

   D. _Trypanosoma cruzi_

   E. _Strongyloides stercoralis_

207. A 12-year-old boy presented from a village in Brazil with itchy papules on the toes and soles of his feet. The lesions were painful and had black dots. A biopsy was obtained from the lesions, and parasite-like structures with exoskeleton and internal parts containing ova were observed in the dermis. What is the likely diagnosis?

   A. Scabies

   B. Tungiasis

   C. Demodicosis

   D. Myiasis

   E. Pediculosis

208. _____ , an insect growth regulator, is used to control flea infestations by preventing hatching of eggs.

   A. Lufenuron

   B. Diazinon

   C. Carbaryl

   D. Rotenone

   E. Deltamethrin

209. _Rhipicephalus_ ticks can transmit _____.

   A. _Babesia bovis_

   B. _Dirofilaria tenuis_

   C. _Toxoplasma gondii_

   D. _Hammondia hammondi_

   E. _Listeria monocytogenes_

210. Which list of *Toxoplasma gondii* life cycle stages is in the right order?

    A. Oocyst, tissue cyst, tachyzoite, gamete, zygote

    B. Zygote, oocyst, gamete, tachyzoite, tissue cyst

    C. Tachyzoite, gamete, tissue cyst, oocyst, zygote

    D. Gamete, tachyzoite, zygote, oocyst, tissue cyst

    E. Tissue cyst, zygote, tachyzoite, oocyst, gamete

211. Which one of the following nematodes has three pointed teeth in the buccal capsule?

    A. *Strongylus vulgaris*

    B. *Strongylus equinus*

    C. *Strongylus edentatus*

    D. *Small Strongyles*

    E. *Strongyloides westeri*

212. In the life cycle of *Taenia solium*, humans play the role of _____.

    A. Definitive host only

    B. Intermediate host only

    C. Definitive host and potentially intermediate host

    D. Transport host

    E. Accidental host

213. _____ serve as the intermediate hosts of *Diphyllobothrium latum*.

    A. Fish

    B. Fleas

    C. Lice

    D. Rabbits

    E. Rodents

214. *Cysticercus pisiformis* can be found in _____.

    A. Cattle

    B. Sheep and goats

C. Buffaloes

D. Dogs and cats

E. Rabbits

215. _____ is the metacestode of *Taenia taenieaformis*.

A. *Cysticercus ovis*

B. *Strobilocercus fasciolaris*

C. *Cysticercus pisiformis*

D. *Cysticercus tenuicollis*

E. *Coenurus cerebralis*

216. _____ is known as large-mouthed bowel worm.

A. *Chabertia ovina*

B. *Trichuris ovis*

C. *Oesophagostomum columbianum*

D. *Oesophagostomum venulosum*

E. *Echinococcus granulosus*

217. _____ is the infective stage of *Trichuris ovis*.

A. First-stage larva (L1)

B. Third-stage larva (L3)

C. Egg containing L1

D. Egg containing L2

E. Egg containing L3

218. The Sellotape test is used for detection of the eggs of _____.

A. *Parascaris equorum*

B. *Strongylus vulgaris*

C. *Oxyuris equi*

D. *Strongylus equinus*

E. *Strongylus edentatus*

219. _____ is the WHO recommended drug to treat leishmaniosis.

    A. Meglumine antimoniate

    B. Pyrimethamine

    C. Sulfadiazine

    D. Spiramycin

    E. Azithromycin

220. Which one of the following parasites causes urogenital schistosomiasis?

    A. *Schistosoma haematobium*

    B. *Austrobilharzia* sp.

    C. *Ornithobilharzia* sp.

    D. *Macrobilharzia* sp.

    E. *Schistosomatium* sp.

221. G6PD-deficient patients should not be treated with _____.

    A. Tafenoquine (TQ) and primaquine (PQ)

    B. Erythromycin

    C. Clindamycin

    D. Doxycycline

    E. Praziquantel

222. An animal disease which can be transmitted to humans is defined as _____.

    A. Zoonosis

    B. Endemic

    C. Enzootic

    D. Anthropogenic

    E. Infectious

223. Which one of the following is the drug recommended for treatment of *Strongyloides stercoralis*?

A. Ivermectin

B. Albendazole

C. Mebendazole

D. Piperazine

E. Miltefosine

224. A parasite that attacks a host other than the usual host is known as _____.

A. Pseudoparasite

B. Obligatory parasite

C. Facultative parasite

D. Accidental parasite

E. Hyperparasite

225. A 40-year-old woman presented with pain in the right upper quadrant (RUQ) of the abdomen. Computed tomography revealed an abscess-like lesion in the hepatobiliary system. A fine-needle transdermal aspiration biopsy of the abscess showed reddish-brown material, containing many microscopic unicellular organisms with pseudopods. What is your diagnosis?

A. Tumour

B. Hydatid cyst

C. Amebic abscess

D. Cysticercosis

E. Cholecystitis

226. A parasite that in all its life cycle stages permanently parasitizes a host, without free living stages, is known as _____.

A. Endoparasite

B. Ectoparasite

C. Temporary parasite

57

D. Permanent parasite

E. Accidental parasite

227. _____ occurs when hypereosinophilic reactions are elicited by larvae passing through the lungs as part of the parasite's developmental cycle in humans.

A. Löffler syndrome

B. Schinzel–Giedion syndrome

C. Nakajo syndrome

D. Parry–Romberg syndrome

E. Sweet's syndrome

228. Which of the following is the correct order of the life cycle stages of hard ticks?

A. Egg, larva, nymph, adult

B. Adult, larva, egg, nymph

C. Larva, nymph, adult, egg

D. Egg, nymph, adult, larva

E. Adult, nymph, egg, larva

229. A 7-year-old girl presents with a history of insomnia, anal pruritus, and scratching of the perianal region especially at night. Blood examination revealed no eosinophilia, anaemia or leucocytosis. Which one of the following parasites is involved in this case?

A. *Trichinella spiralis*

B. *Necator americanus*

C. *Enterobius vermicularis*

D. *Ancylostoma duodenale*

E. *Ascaris lumbricoides*

230. In congenital toxoplasmosis, folic acid is prescribed to _____.

A. Antagonize the action of pyrimethamine

B. Potentiate the action of pyrimethamine

C. Mitigate the haematological toxicities of pyrimethamine

D. Support the patient immune defenses

E. Protect against opportunistic viruses

231. Relapses occur even years after the first infection in which of the following *Plasmodium* sp.?

A. *Plasmodium falciparum*

B. *Plasmodium ovale*

C. *Plasmodium vivax*

D. *Plasmodium knowlesi*

E. *Plasmodium malariae*

232. To eliminate liver stages of *P. vivax* that would drive recurrence, patients should take _____ drug.

A. Primaquine

B. Chloroquine

C. Quinine

D. Artemisinin-based combination therapies

E. Flucloxacillin

233. Members of which of the following superfamilies rely on multiple hosts in their life cycle (i.e. heteroxenous)?

A. Ancylostomatoidea

B. Diaphanocephaloidea

C. Strongyloidea

D. Metastrongyloidea

E. Trichostrongyloidea

234. _____ is the most predominant *Toxoplasma gondii* genotype in mainland China.

A. ToxoDB#2

B. ToxoDB#9

C. ToxoDB#10

D. ToxoDB#19

E. ToxoDB#52

235. Which one of the following parasites is associated with 'measly pork'?

    A. *Diphyllobothrium latum*

    B. *Dipylidium caninum*

    C. *Taenia saginata*

    D. *Taenia solium*

    E. *Trichinella spiralis*

236. A teenage boy presented with a 14-day history of itching in his groin. On dermoscopic investigation, there were moving, crab-like insects, and nits were observed attached to his pubic hair. What is your diagnosis?

    A. Infestation with *Pediculus humanus capitis*

    B. Infestation with *Tunga penetrans*

    C. Infestation with *Phthirus pubis*

    D. Infestation with *Pediculus humanus corporis*

    E. Infestation with *Pulex irritans*

237. Which list of phases of *Cryptosporidium* life cycle is in the right sequence?

    A. Sporozoite, trophozoite, meront, oocyst, zygote, merozoite, gamete

    B. Sporozoite, oocyst, trophozoite, gamete, meront, merozoite, zygote

    C. Oocyst, sporozoite, trophozoite, meront, merozoite, gamete, zygote

    D. Zygote, trophozoite, merozoite, gamete, meront, oocyst, sporozoite

    E. Trophozoite, gamete, merozoite, oocyst, meront, zygote, sporozoite

238. Aside from finding adult fleas on a cat or dog, the presence of flea dirt is also considered a definitive diagnosis of an infestation. What is 'flea dirt'?

    A. Flea faeces

    B. Flea urine

    C. Flea bite

    D. Oral secretion of fleas

    E. Flea cuticle

239. A 25-year-old man from Italy presented with irritation and itching in his right eye. Also, three white, slender, and filiform round-worms were recovered via irrigation of the conjunctival sac. These worms measured ~1 cm long and had striations over the entire tegumental surface. On the basis of these findings, what is the most likely diagnosis?

    A. Gnathostomiasis

    B. Thelaziosis

    C. Trichinosis

    D. Toxocariasis

    E. Onchocerciasis

240. Which one of the following *Plasmodium* stages is ingested by a mosquito during blood meal?

    A. Sporozoite

    B. Merozoites

    C. Gametes

    D. Ookinete

    E. Hyponozoite

241. All of the following are correct about *Macracanthorhynchus hirudinaceus* except?

    A. It is the giant thorny-headed worm of the pig

    B. Adult worms are found within renal tissue

    C. Its larval stage is known as acanthor

    D. Beetles serve as intermediate hosts

    E. It is an acanthocephalan worm

242. Prevention of bovine lungworm is best achieved by _____.

    A. Regular anthelmintics

    B. Annual vaccination

    C. Pasture rotation

    D. Alternate grazing with sheep

    E. Avoiding or draining snail habitats

243. Which of the following drugs does not match the chemical class?

A. Heterocyclic compound: piperazine

B. Benzimidazole: Zolvix

C. Macrocyclic lactone: Moxidectin

D. Isoquinolone: Praziquantel

E. Tetrahydropyrimidine: Pyrantel

244. Which one of the following tapeworms is a pseudophyllidean?

A. *Echinococcus granulosus*

B. *Taenia saginata*

C. *Diphyllobothrium latum*

D. *Dipylidium caninum*

E. *Taenia solium*

245. A growing body of evidence suggests a correlation between schizophrenia and chronic infection with _____.

A. *Acanthamoeba castellani*

B. *Toxoplasma gondii*

C. *Neospora caninum*

D. *Leishmania donovani*

E. *Naegleria fowleri*

246. A 12-year-old boy from Mexico presented complaining of mobile pruritic eruption on his left foot. Physical examination of the affected foot revealed a serpiginous, erythematous raised tract. What is the most likely diagnosis?

A. Cutaneous larva migrans

B. Dirofilariasis

C. Scabies

D. Pediculosis

E. Dermatophytosis

247. Onchosphere is the larval form of which of the following parasites?

    A. *Taenia solium*

    B. *Toxocara vitulorum*

    C. *Diphyllobothrium latum*

    D. *Ascaris lumbricoides*

    E. *Fascioloides magna*

248. Which one of the following parasites can cause hyperinfection or disseminated infection?

    A. *Ascaris lumbricoides*

    B. *Ancylostoma duodenale*

    C. *Strongyloides stercoralis*

    D. *Trichuris trichiura*

    E. *Enterobius vermicularis*

249. Which one of the following parasites glues its eggs/nits to the hair shaft of the host?

    A. Fleas

    B. Ticks

    C. Lice

    D. Mites

    E. Bugs

250. _____ mites are identified by their three-segmented pedicels and funnel-shaped suckers.

    A. *Sarcoptes scabiei*

    B. *Psoroptes ovis*

    C. *Chorioptes bovis*

    D. *Dermanyssus gallinae*

    E. *Otedectes cynotis*

Answers: Multiple Choice Questions

1.	A	29.	D
2.	C	30.	C
3.	C	31.	E
4.	A	32.	C
5.	E	33.	B
6.	A	34.	C
7.	A	35.	C
8.	E	36.	A
9.	C	37.	C
10.	C	38.	B
11.	B	39.	A
12.	A	40.	B
13.	E	41.	B
14.	B	42.	A
15.	C	43.	B
16.	A	44.	D
17.	A	45.	C
18.	E	46.	C
19.	A	47.	D
20.	A	48.	A
21.	A	49.	A
22.	C	50.	A
23.	B	51.	C
24.	D	52.	A
25.	A	53.	A
26.	C	54.	A
27.	A	55.	B
28.	E	56.	A

57. A		86. B	
58. A		87. D	
59. A		88. E	
60. E		89. A	
61. B		90. A	
62. B		91. D	
63. E		92. A	
64. E		93. B	
65. A		94. A	
66. E		95. A	
67. B		96. A	
68. C		97. E	
69. A		98. B	
70. C		99. C	
71. B		100. A	
72. A		101. E	
73. C		102. C	
74. C		103. A	
75. B		104. C	
76. B		105. A	
77. A		106. D	
78. E		107. A	
79. D		108. C	
80. A		109. A	
81. A		110. C	
82. B		111. A	
83. B		112. A	
84. A		113. A	
85. A		114. A	

115.	A	144.	B
116.	B	145.	A
117.	A	146.	A
118.	B	147.	A
119.	A	148.	C
120.	A	149.	A
121.	A	150.	E
122.	A	151.	E
123.	A	152.	A
124.	B	153.	A
125.	B	154.	D
126.	A	155.	C
127.	E	156.	E
128.	B	157.	A
129.	C	158.	A
130.	D	159.	A
131.	B	160.	A
132.	A	161.	C
133.	E	162.	C
134.	A	163.	B
135.	B	164.	C
136.	E	165.	D
137.	B	166.	C
138.	A	167.	C
139.	C	168.	B
140.	B	169.	B
141.	A	170.	A
142.	A	171.	A
143.	A	172.	A

173. A		202. A	
174. D		203. A	
175. C		204. B	
176. C		205. A	
177. A		206. A	
178. B		207. B	
179. B		208. A	
180. C		209. A	
181. A		210. A	
182. A		211. B	
183. D		212. C	
184. A		213. A	
185. A		214. E	
186. A		215. B	
187. A		216. A	
188. E		217. C	
189. A		218. C	
190. A		219. A	
191. A		220. A	
192. E		221. A	
193. A		222. A	
194. A		223. A	
195. B		224. D	
196. A		225. C	
197. A		226. D	
198. A		227. A	
199. B		228. A	
200. C		229. C	
201. A		230. C	

231. C	241. B
232. A	242. B
233. D	243. B
234. B	244. C
235. D	245. B
236. C	246. A
237. C	247. A
238. A	248. C
239. B	249. C
240. C	250. B

2 Matching Questions

PART 1 Extended Matching Questions

(**I**) The text box contains a list of **diseases.**

A. Babesiosis	F. Haemonchosis
B. Borreliosis	G. Toxocariasis
C. Leishmaniosis	H. Dirofilariosis
D. Giardiosis	I. Coccidiosis
E. Mal de cadeiras	J. Cryptosporidiosis

For each sentence, select the disease that is most likely being described.

1. _____ is caused by a spirochete (corkscrew-shaped bacteria) spread by ticks.

2. _____ is caused by a protozoan which lives in the erythrocytes and is spread between hosts by ticks.

3. _____ is caused by a protozoan which is spread between hosts by sandflies.

4. _____ is caused by roundworms of dogs and cats. The affected persons may exhibit abdominal discomfort, cough, wheezing, or even loss of vision.

5. _____ is caused by *Trypanosoma evansi*. Affected horses can exhibit ataxia, head tilt and circling, head pressing, hyperexcitability, and staggering gait.

6. _____ disease, according to the WHO 2016 report, causes 18% of the deaths in children under 5 years old.

7. _____ is a mosquito-borne disease characterized by the presence of *Dirofilaria immitis* worms in the pulmonary arteries.

8. _____ is caused by protozoan parasites (*Eimeria* spp.) that cause decreased growth and feed efficiency, and mortality in floor-raised chickens.

9. _____ is an enteric protozoal disease that is often seen in international travellers and daycare workers.

10. _____ is one of the most economically significant parasitic diseases of small ruminants worldwide. It is caused by haematophagous nematodes knowns as 'Barber's pole worm'.

(**II**) The text box contains a list of **drugs**.

A. Ivermectin	F. Fipronil
B. Advocate	G. Imidacloprid
C. Melarsomine dihydrochloride	H. Metronidazole
D. Praziquantel	I. Permethrin
E. Artemisinin combination	J. Pyrantel

For each sentence, select the drug that is most likely being described.

1. _____ does not affect trematodes and cestodes because these groups of flatworms lack GABA receptors.

2. _____ is also used in a wide range of flea treatment, such as spot-on solutions and a slow-release collar.

3. Oral _____ is the recommended treatment for the management of trichomoniasis.

4. _____ is indicated for monthly use for the prevention of canine angiostrongylosis and patent infection, with 100% efficacy against L4 larvae and immature adults (L5) of *Angiostrongylus vasorum*.

5. _____ is the only FDA-approved product for heartworm adulticidal therapy.

6. _____ is effective against both adult and juvenile forms of tapeworm. It is the treatment of choice where routine chemical prophylaxis against echinococcosis is required.

7. _____ has high safety margins in cats and dogs but should not be used in rabbits.

8. _____ is safe at therapeutic doses, but cats do not tolerate therapeutic doses for dogs.

9. _____ should not be used concurrently with piperazine.

10. _____ is the drug of choice in the treatment of uncomplicated *Plasmodium falciparum* malaria in most endemic countries.

(III) The text box contains a list of **diseases**.

A. Onchocerciasis	F. Scabies
B. Lyme disease	G. Myiasis
C. Leishmaniosis	H. Cheyletiellosis
D. Balantidiasis	I. Otodectic mange
E. Trombiculosis	J. Filariasis

For each sentence, select the disease that is most likely being described.

1. _____ is known as 'river blindness' because it is restricted to areas adjacent to river systems.

2. _____ is generally asymptomatic, but diarrhoea and abdominal aches may occur. A few patients may develop intestinal perforation or dysentery associated with haemorrhage, resembling amoebic dysentery.

3. _____ is known as 'kala-azar', and is associated with enlargement of the spleen and liver, and anaemia.

4. _____ is a tick-borne disease caused by *Borrelia burgdorferi*, and can lead to the development of dermatological, neurological, arthritic, and cardiac manifestations.

5. _____ is an infestation caused by larval stage of mites, also known as 'chiggers'.

6. _____ is a contagious skin infestation with *Sarcoptes scabiei* mites, characterized by thick crust on the skin.

7. _____ is an invasion of animal or human tissue with maggots of several fly species.

8. _____ is a skin infestation caused by mites, which cause affected cats to exhibit head shaking, ear scratching, aural pruritus, and ear droop.

9. _____ is an arthropod-borne infection caused by tissue-invasive roundworms, namely *Brugia malayi* and *B. timori*, *Wuchereria bancrofti*, *Onchocerca volvulus*, and *Loa loa*.

10. _____ is a skin disease caused by non-burrowing mites known as 'walking dandruff' because they look like white, mobile dandruff flakes.

(IV) The text box contains a list of **parasite agents**.

A. *Leishmania* sp.	F. *Cryptosporidium*
B. *Trypanosoma equiperdum*	G. *Paramphistomum*
C. *Dicrocoelium dendriticum*	H. *Strongyloides stercoralis*
D. *Gongylonema* sp.	I. *Triatoma* spp.
E. *Schistosoma*	J. *Demodex* spp.

For each sentence, select the parasite species that is most likely being described.

1. _____ is a digenetic protozoan parasite, which cycles between a flagellated promastigote stage in the gastrointestinal tract of sandflies and an intracellular amastigote stage in the mammalian macrophages.

2. Following ingestion of metacercariae, _____ excyst in the duodenum and migrate upstream to become adult flukes in the rumen and reticulum.

3. _____ is a small lancet-like fluke that infects sheep and cattle. Land snails and ants serve as the first and second intermediate hosts, respectively.

4. In humans, _____ is the only known nematode to reside in the submucosa of the oral cavity and oropharynx.

5. _____ is the causative agent of dourine or 'mal du coit'.

6. Female _____ worms mature through pairing with the male worm after travelling through the circulatory system.

7. _____ is the second most prevalent cause of moderate-to-severe diarrhoea in children less than 2 years old.

8. _____ has a complex life cycle, involving both free-living and parasitic scenarios.

9. _____ are known as 'kissing bugs' and serve as vectors of *Trypanosoma cruzi* – the causative agent of Chagas disease.

10. _____ are known as 'follicular mites' because they inhabit the hair follicles of the affected host.

(V) The text box contains a list of **diagnostic tests**.

A. The Angio Detect™ Test	F. 'Sticky-tape' impression
B. McMaster technique	G. Xenodiagnosis
C. Direct faecal smear	H. PCR
D. Baermann technique	I. Larval culture
E. Bronchoalveolar lavage	J. Skin scraping

For each sentence, select the diagnostic test that is most likely being described.

1. _____ is a pet-side antigen blood test, which is used for the detection of *Angiostrongylus vasorum* infection in dogs.

2. _____ is used for isolation of larvae from faeces.

3. _____ is used for counting the number of nematode ova per gram of faeces.

4. _____ provides a quick and simple but relatively insensitive method for demonstrating helminth infection and identifying the eggs and larvae present.

5. _____ is used to identify the third-stage larvae of nematodes present in faeces.

6. _____ is used for detecting *Oxyuris equi* eggs, which aren't excreted in faeces.

7. _____ employs the use of laboratory-based arthropod vectors to detect low levels of parasites in infected hosts.

8. _____ involves the isolation of nucleic acid and amplification of a portion of genetic material of the agent of interest.

9. _____ is used for the detection of ectoparasites, such as mites, in the skin of the affected host.

10. _____ is used for the diagnosis of lungworm infection and is done by inserting a tube into the larynx and bronchi.

PART 2 Cross Matching Questions

(**I**) Pair items in column A to the most relevant items in column B.

Column A	Column B
1. *Haemonchus contortus*	A. An important cause of abortion in sheep
2. 'Bottle jaw' in sheep	B. Hexagonal basis capitulum and bifid first coxa in both sexes
3. Tapeworms	C. Has a genal comb (ctenidium) with a row of 4–6 vertical spines
4. *Toxoplasma gondii*	D. Vector of liver flukes
5. *Dioctophyma renale*	E. Eggs contain oncosphere
6. *Trichodectes canis*	F. Warble tumour-like swelling in the skin
7. *Rhipicephalus sanguineus*	G. An intermediate host for the flea tapeworm
8. *Lymnaea* spp.	H. Giant kidney worm
9. *Spilopsyllus cuniculi*	I. Pallor of the mouth mucous membrane
10. *Hypoderma bovis*	J. Associated with chronic hypoalbuminaemia

Write the correct letter next to the figures 1–10.

1.   6.

2.   7.

3.   8.

4.   9.

5.   10.

(II) Pair items in column A to the most relevant items in column B.

Column A	Column B
1. Tromiculids	A. Lives in the pulmonary arteries of rats
2. Hydatid sand	B. Nasal bot of sheep
3. *Ixodes* spp.	C. Causes 'white-spot disease'
4. *Cryptosporidium*	D. Only parasitic in the larval stage
5. *Angiostrongylus cantonensis*	E. Protoscolices of *Echinococcus granulosus*
6. *Ctenocephalides canis*	F. Leads to esophageal neoplasms
7. *Spirocerca lupi*	G. Anal groove is anterior to the anus
8. *Stephanurus dentatus*	H. Although intracellular, it does not invade the cell cytoplasm
9. *Oestrus ovis*	I. On the genal comb (ctenidium), the first spine is half as long as the second spine
10. *Ichthyophthirius multifiliis*	J. Lives in the kidneys and perirenal tissue of pigs

Write the correct letter next to the figures 1–10.

1.   6.

2.   7.

3.   8.

4.   9.

5.   10.

75

(III) Pair items in column A to the most relevant items in column B.

Column A	Column B
1. Canker	A. Non-burrowing red poultry mite
2. *Sarcocystis neurona*	B. Infest the gills and skin of fish
3. *Oslerus osleri*	C. Ingestion of raw or improperly cooked fish
4. Hypobiosis	D. Conical flukes with both anterior and posterior suckers
5. Biting midges	E. Found in nodules adjacent to the tracheal bifurcation
6. *Dermanyssus gallinae*	F. Transmitted in crop milk when the parent pigeons feed their chicks
7. Paramphistome	G. Human acanthocephaliasis
8. *Dactylogyrus* spp.	H. Arrested or inhibited development of larvae
9. *Anisakis simplex*	I. Agent of equine protozoal myeloencephalitis
10. *Moniliformis moniliformis*	J. Insect hypersensitivity (sweet itch)

Write the correct letter next to the figures 1–10.

1.   6.
2.   7.
3.   8.
4.   9.
5.   10.

(IV) Pair items in column A to the most relevant items in column B.

Column A	Column B
1. *Cysticercus tenuicollis*	A. Culicoides
2. Occult heartworm infection	B. Obtuse angle between pairs of merozoites
3. *Babesia gibsoni*	C. A subcutaneous dwelling filarial nematode of dogs
4. Trypanosome	D. Occurs in the peritoneum and liver of sheep
5. Fish louse	E. Without circulating microfilariae
6. *Acanthocheilonema reconditum*	F. Eggs are not shed in the faeces of the affected host
7. *Parelaphostrongylus tenuis*	G. Leaf-like protozoan with a single flagellum

Column A	Column B
8. *Capillaria* spp.	H. Ileal impaction and spasmodic colic in the affected horse
9. Sweet itch	I. Aberrant migration within the CNS
10. *Anoplocephala perfoliata*	J. A branchiuran parasite of the genus *Argulus*

Write the correct letter next to the figures 1–10.

1.   6.
2.   7.
3.   8.
4.   9.
5.   10.

(V) Pair items in column A to the most relevant items in column B.

Column A	Column B
1. *Babesia bigemina*	A. Lung fluke of dogs, cats, mink, pigs, and wild carnivores
2. *Spirometra erinacei*	B. A soil saprophyte which aberrantly infects horses
3. *Amoebotaenia*	C. Produces 'alveolar' cysts
4. *Stilesia and Thysanosoma*	D. Have two longitudinal grooves (bothria) on both sides of the scolex
5. *Psorergates ovis*	E. Aspirates may be examined for detection of trophozoites
6. *Dipylidium caninum*	F. Acute angle between pairs of merozoites
7. *Echinococcus multilocularis*	G. Found in the small intestine of poultry
8. *Paragonimus kellicotti*	H. Found in the bile ducts of ruminants
9. *Halicephalobus gingivalis*	I. 'Itch mite' of sheep
10. *Giardia duodenal*	J. Gravid proglottids resemble 'cucumber seeds'

Write the correct letter next to the figures 1–10.

1.   6.
2.   7.
3.   8.
4.   9.
5.   10.

Answers: Matching Questions

PART 1 Extended Matching Questions

(I)	(II)	(III)	(IV)	(V)
1. B	1. A	1. A	1. A	1. A
2. A	2. G	2. D	2. G	2. D
3. C	3. H	3. C	3. C	3. B
4. G	4. B	4. B	4. D	4. C
5. E	5. C	5. E	5. B	5. I
6. J	6. D	6. F	6. E	6. F
7. H	7. F	7. G	7. F	7. G
8. I	8. I	8. I	8. H	8. H
9. D	9. J	9. J	9. I	9. J
10. F	10. E	10. H	10. J	10. E

PART 2 Cross Matching Questions

(I)	(II)	(III)	(IV)	(V)
1. I	1. D	1. F	1. D	1. F
2. J	2. E	2. I	2. E	2. D
3. E	3. G	3. E	3. B	3. G
4. A	4. H	4. H	4. G	4. H
5. H	5. A	5. J	5. J	5. I
6. G	6. I	6. A	6. C	6. J
7. B	7. F	7. D	7. I	7. C
8. D	8. J	8. B	8. F	8. A
9. C	9. B	9. C	9. A	9. B
10. F	10. C	10. G	10. H	10. E

3 Fill-in-the-Blank Questions

1. Humans become infected with *Toxocara* sp. when they accidentally ingest _____ in soil contaminated by dog or cat faeces.

2. The female schistosome resides inside a specialized canal, on the ventral surface of the male, known as _____.

3. Gamma amino butyric acid (GABA) acts as an inhibitory neurotransmitter and inhibits _____ in nematodes.

4. _____ are the only known definitive hosts for *Toxoplasma gondii*.

5. The most common *Sarcocystis* infection of horses is equine protozoal myeloencephalitis (EPM) which is caused by _____.

6. *Sarcocystis* was originally reported in 1843 as white thread-like structures (_____) in the muscle of mice.

7. Chagas disease is caused by _____, which is transmitted by *Triatominae* insects, predominantly in rural areas of Latin America.

8. _____ intracellular bacteria are essential for development of larvae and the survival of adult *Onchocerca volvulus* worms.

9. _____ was identified in 1836 when motile organisms were observed in the vaginal fluid.

10. _____ was discovered in 1909 by the English protozoologist Charles Wenyon after examining his own parasitological stool preparations.

11. _____ tapeworm can be identified based on the presence of double sets of reproductive organs in each segment/proglottid with a genital pore on both lateral edges.

12. _____ mites have dorsal striations and a dorsally located anus compared with the terminal anus, dorsal pegs and spines found in *Sarcoptes* mites.

13. Field diagnosis of *Eimeria* infection in poultry depends on the identification of _____ and microscopic examination of faecal droppings and litter for _____.

14. Larval development of the nematode _____, from egg to the L3 stage, occurs within the eggshell.

15. Out of the 13 *Eimeria* species reported to infect cattle, _____ and _____ are the most pathogenic species, leading to typhlocolitis associated with bloody diarrhoea and dehydration, especially in calves.

16. Cryptosporidiosis in humans is mainly caused by _____ (zoonotic) and _____ (human specific).

17. _____ is a serious disease of salmonids caused by the myxosporean parasite _____, which has a two-host life cycle, utilizing _____ as an intermediate host.

18. *Toxoplasma gondii* infection in immunocompetent individuals is generally _____. However, a small proportion of acutely infected patients may develop cervical _____ or a _____.

19. The intestinal fluke _____ was originally described by George Busk in 1843 from an Indian sailor.

20. Six *Entamoeba* species have been described in humans. These include _____, _____, _____, _____, _____, and _____. Among these, _____ is the only pathogenic species.

21. Liver flukes can be treated with the flukicide _____ or _____.

22. _____ protozoa live in intracellular but extracytosolic parasitophorous vacuoles beneath the plasma membrane of intestinal epithelial cells.

23. _____ is an apicomplexan protozoan parasite that is capable of infecting humans, and almost all species of animals and birds.

24. Infection with _____ was first recognized by Babès in 1888 inside red blood cells of cattle in Romania.

25. The completion of the life cycle of *Babesia* requires a mammalian host and _____ ticks.

26. About 12 *Eimeria* species are reported to infect the intestinal tract of the rabbits, except one species, _____, which infects the liver.

27. The three principal *Schistosoma* species infecting humans are _____, _____, and _____.

28. Imidacloprid activity is attributed to its effect on the postsynaptic _____ in the CNS of the insect.

29. _____ (i.e. cutaneous habronemiasis) are pruritic, pyogranulomatous lesions caused by aberrant cutaneous migration of the larvae of the equine stomach worm, _____.

30. _____ is the tapeworm of foxes, dogs, and to some extent cats (definitive hosts), with rodents as intermediate hosts.

31. The life cycle of _____ involves a rodent definitive host and a mollusk intermediate host, often a slug or land snail.

32. The blood flukes schistosomes have a life cycle that involves freshwater _____ as an intermediate host and a vertebrate definitive host.

33. Human African trypanosomiasis is caused by infection with the tsetse fly-transmitted haemoflagellates _____ and _____.

34. Three primary lungworms of small ruminants are of clinical and economic importance: _____, _____ and _____.

35. Damage inflicted by liver flukes may predispose the affected host to invasion by anaerobic _____ such as *Clostridium novyi*, which may lead to fatal disease, which is known as _____.

36. *Taenia crassiceps* differs from other tapeworms because of its ability to bud _____ inside the host and generate _____ outside the cyst.

37. Adult *Diphyllobothrium latum* cause limited clinical illness, however they compete with their host for absorption of _____ from the intestinal lumen, resulting in _____ anaemia.

38. The _____ drugs are effective against both endo- and ectoparasites; they are known by the term 'endectocides'.

39. _____ is the first-choice drug for treatment of early Lyme disease because it is effective for treatment of erythema migrans and human granulocytic anaplasmosis, which may occur simultaneously with early Lyme disease.

40. _____ is an agonist of invertebrate-specific, glutamate-activated, inhibitory chloride channels; it causes flaccid paralysis and subsequent death in nematodes and mites.

41. _____ and _____ are possible adverse effects in dogs and cats treated with Albendazole.

42. The _____ system is a tool used for assessment of the extent of anaemia caused by *Haemonchus contortus* infection, and has been developed for the identification of those animals who require treatment in the sheep flock.

43. The Chinese liver fluke disease 'clonorchiasis' is caused by _____. Humans are infected by eating raw, undercooked or salted fish containing _____.

44. *Neospora caninum* is a cyst-forming coccidian with an indirect life cycle that involves _____ as the definitive host and _____ as the main intermediate host.

45. The striking feature in dogs with *Dipylidium caninum* is the passage of motile _____, which can be found in the perianal region and in the freshly voided faeces.

46. *Sarcocystis hominis* is mildly pathogenic for humans and is acquired by ingestion of _____.

47. _____ and _____ are the two most pathogenic coccidial species that affect the intestine of rabbits.

48. In addition to *Sarcocystis neurona*, _____ is another intracellular coccidian protozoan that has been involved to a lesser extent in causing equine protozoal myeloencephalitis (EPM).

49. Fipronil is a member of the insecticide class _____. It is used on cats and dogs to treat _____, and all stages of ticks and mites.

50. Human _____, is the direct result of ingesting *Trichinella* larvae in _____.

51. When humans infected with *Taenia solium* serve as an intermediate host, they can develop a serious condition known as _____.

52. *Taenia crassiceps* has a life cycle that involves _____, _____, and _____ as the final hosts and rodents as _____.

53. The life cycle of *Toxoplasma gondii* encompasses 3 stages: _____, which rapidly proliferate and cause acute infection; _____, which remain dormant inside tissue cysts; and sporozoites inside _____.

54. _____, the beef tapeworm, and *Taenia solium*, the pork tapeworm, are the only *Taenia* spp. affecting humans.

55. _____ was the first anthelmintic to be developed and licensed at the beginning of the 1960s.

56. Humans acquire *Taenia saginata* infection by eating _____ infected with the cysticerci.

57. Of the 200+ species of *Sarcocystis*, humans serve as the definitive host for two (_____ and _____).

58. Cattle develop _____ by grazing pasture upon which human faeces has been deposited.

59. Multiple tapeworms inhabit the small intestine of dogs. The most prevalent species is _____. Other species include _____, _____, _____, _____, and _____.

60. Definitive diagnosis of *Dipylidium caninum* is achieved via identification of _____ or _____ on the faeces or around the anus, respectively.

61. The mouse is the intermediate host of the cat tapeworm, _____. The metacestode of this parasite is given its own scientific name, _____, and can be found in the _____ and _____ of infected mice ~30 days after ingestion of eggs.

62. Amitraz is an acaricide used in the control of _____, _____, and _____ on domestic animals.

63. Treatment of cattle infected with *Hypoderma* while the larvae are in the spinal canal may cause _____.

64. _____ and _____ are surgical procedures used by some sheep breeders to reduce the risk of flystrike in sheep.

65. Treatment of dogs with _____ requires mechanical removal of a large number of worms using grasping forceps.

66. Excretory-secretory products released by helminth parasites are efficient inducers of _____ of immune response.

67. _____ is a skin condition in dogs and cats that are sensitized to flea saliva proteins through flea bites.

68. *Cimex lectularius*, the common _____, hide in crevices and cracks in walls, mattresses, furniture, or under carpeting. These insects are usually only active during the night.

69. The raccoon nematode, _____, can cause _____ in humans, especially children.

70. Amitraz is a centrally acting _____ agonist that also inhibits _____ and _____ synthesis.

71. The sheep ked, _____, are wingless blood-sucking insects, which spend most of their life on sheep.

72. Both tetrahydropyrimidines and imidazothiazoles act as _____ receptor agonists.

73. Parasites can be referred to as _____ when they are free-living environmental stages or living inside the animals but have not been exposed to anthelmintic treatment.

74. Adulticide treatment in canine patients is often associated with serious complications, such as _____.

75. A characteristic morphological feature of *Haematopinus suis* is the presence of _____ on the sides of each abdominal segment.

76. Larvae of warble flies of the genus *Hypoderma* migrate throughout the flesh of the affected cattle creating tracks of greenish gelatinous material known as '_____'.

77. Parasites that have been identified in the brain or spinal cord of horses include protozoa (equine protozoal myeloencephalitis caused by _____, _____), rhabditid nematodes (_____), strongyloid nematodes (_____, _____, _____, _____), spiruroid nematodes (_____), filarid nematodes (_____ spp.), and warble fly larvae (_____ spp.).

78. The suboptimal dosing of anthelmintics can lead to the development of _____.

79. The free-living L3s of *Strongyloides* parasites can be acquired from the environment via _____ and _____ invasion.

80. Human sparganosis occurs by drinking water contaminated with _____ harbouring _____ larvae.

81. Morphological differentiation between male and female can be achieved easily in _____ lice compared to _____ lice.

82. _____ is caused by a toxin secreted into a host through saliva from the salivary glands of feeding female ticks. Small children and individuals exposed to tick-endemic areas are at high risk particularly during the peak tick seasons in _____ and _____.

83. _____ are innate immune receptors, whose function is to recognize pathogen-associated molecular patterns, and induce innate and adaptive immune responses to counter the infection.

84. In recent years, the use of _____ has been presented as an eco-friendly biological means for control of free-larval stages of gastrointestinal worms in small ruminants.

85. The aberrant migration of the _____ in alpacas and llamas is associated with a neurological disease known as cerebrospinal nematodiasis.

86. _____ is an elongated tongue-shaped pentastomid parasite that can be found in the nasal cavities or sinuses of dogs and foxes.

87. The damage caused by *Oxyuris equi* infection in horses is attributed to _____.

88. Benzimidazoles work by binding to _____, which inhibits _____.

89. _____ is a condition produced by parasitic larvae of flies (e.g. *Oestrus ovis*, *Hypoderma bovis*) that are dependent on a host for part of their life cycle, and in which larvae colonize undamaged vital tissues, such as skin.

90. The use of blowfly larvae to debride, disinfect, and promote healing of soft tissue wounds is known as _____.

91. *Trypanosoma brucei* brucei causes _____ in cattle, and *T. brucei* rhodesiense and *T. brucei* gambiense cause _____ in humans.

92. _____ is the only *Trypanosoma* not transmitted by insects, but instead is transmitted from infected to healthy animals during sexual intercourse.

93. Sheep blowflies thrive in a climate which is _____ and _____.

94. The hind limb weakness in horses infected with the flagellate protozoan *Trypanosoma evansi* leads to the disease named '_____' in Brazil or '_____' in Spanish-speaking countries of South America.

95. The sandfly inoculates promastigotes into the skin of the mammalian host, which are subsequently engulfed by reticulo-endothelial cells and transform into intracellular _____ form.

96. _____, _____, _____, and _____ are the scientific name of spotted flesh fly, sheep nasal bot, old world screwworm fly, and rodent botfly, respectively.

97. *Cryptosporidium* causes serious waterborne outbreaks due to the resistance of the parasite to _____ or other disinfectants.

98. Among the five *Plasmodium* species that cause malaria in humans (_____, _____, _____, _____, and _____), _____ is the most virulent and causes most of the deaths.

99. *Plasmodium vivax* infection in humans consists of blood (_____) and _____ (exoerythrocytic) stages.

100. Several intracellular bacteria species, such as _____, _____, _____, *Chlamydia*, and _____, can survive within protozoa, such as *Acanthamoeba* spp.

Answers: Fill-in-the-Blank Questions

1. infectious *Toxocara* eggs containing second larval stage.
2. Gynecophoric canal.
3. post-synaptic stimulation of the neurons.
4. Members of the family Felidae.
5. *Sarcocystis neurona*.
6. known as Meischer's tubules.
7. *Trypanosoma cruzi*.
8. *Wolbachia*.
9. *Trichomonas vaginalis*.
10. *Dientamoeba fragilis*.
11. *Dipylidium caninum*.
12. *Notoedres*.
13. intestinal pathological lesions; *Eimeria* oocysts.
14. *Nematodirus*.
15. *E. zuernii*; *E. bovis*.
16. *Cryptosporidium parvum*; *Cryptosporidium hominis*.
17. Whirling disease; *Myxobolus cerebralis*; *Tubifex tubifex*.
18. Asymptomatic; lymphadenopathy; flu-like illness.
19. *Fasciolopsis buski*.
20. *Entamoeba histolytica*; *E. dispar*; *E. moshkovskii*; *E. polecki*; *E. coli*; *E. hartmanii*; *E. histolytica*.
21. clorsulon; albendazole.
22. *Cryptosporidium*.
23. *Toxoplasma gondii*.
24. *Babesia*.
25. *Ixodes*.
26. *Eimeria stiedae*.
27. *Schistosoma mansoni*; *S. japonicum*; *S. haematobium*.
28. nicotinic acetylcholine receptors (nAChRs).
29. Summer sores; *Habronema*.
30. *Echinococcus multilocularis*.
31. *Angiostrongylus cantonensis*.
32. snails.
33. *Trypanosoma brucei* rhodesiense; *Trypanosoma brucei* gambiense.
34. *Dictyocaulus filaria*; *Muellerius capillaris*; *Protostrongylus rufescens*.
35. *Clostridium*; bacillary haemoglobinuria.
36. exogenously; new buds.
37. vitamin B12; megaloblastic.
38. macrocyclic lactone.
39. Doxycycline.
40. Ivermectin.
41. Leukopenia; thrombocytopenia.
42. FAMACHA.

43. *Clonorchis sinensis*; metacercariae.
44. Dogs; cattle.
45. proglottids.
46. uncooked beef containing sarcocysts.
47. *Eimeria magna*; *Eimeria irresidua*.
48. *Neospora hughesi*.
49. Phenylpyrazoles; fleas.
50. trichinellosis; raw or undercooked pork.
51. cysticercosis.
52. Foxes; dogs; felids; intermediate hosts.
53. Tachyzoites; bradyzoites; oocysts.
54. *Taenia saginata*.
55. Thiabendazole.
56. raw or insufficiently cooked beef.
57. *Sarcocystis hominis*; *Sarcocystis suihominis*.
58. *Cysticercus bovis*.
59. *Dipylidium caninum*; *Taenia pisiformis*; *Echinococcus granulosus*; *Taenia multiceps*; *Mesocestoides* sp.; *Spirometra* sp.
60. egg capsules; proglottids.
61. *Taenia taenieaformis*; *Cysticercus fasciolaris*; muscle; liver.
62. ticks; mites; lice.
63. paraplegia.
64. Tail docking; mulesing.
65. Caval syndrome.
66. Th2 type.
67. Flea allergy dermatitis.
68. bed bugs.
69. *Baylisascaris procyonis*; neural larva migrans.
70. α-2 adrenergic; monoamine oxidase; prostaglandin.
71. *Melophagus ovinus*.
72. nicotinic acetylcholine.
73. in refugia.
74. Pulmonary thromboembolism.
75. paratergal plates.
76. butcher's jelly.
77. *Sarcocystis neurona*; *Neospora hughesi*; *Halicephalobus gingivalis*; *Strongylus vulgaris*; *S. equinus*; *Angiostrongylus cantonensis*; *Parelaphostrongylus tenuis*; *Draschia megastoma*; *Setaria*; *Hypoderma*
78. anthelmintic resistance.
79. oral ingestion; transcutaneous.
80. copepods; procercoid.
81. sucking; biting.
82. Tick paralysis; spring; summer.
83. Toll-like receptors.

84. nematode-trapping fungi.
85. meningeal worm.
86. *Linguatula serrata.*
87. egg-laying behaviour of the females.
88. β-subunit of tubulin; microtubule formation.
89. Specific (also known as obligatory) myiasis.
90. maggot debridement therapy.
91. nagana; sleeping sickness.
92. *Trypanosoma equiperdum.*
93. warm; humid.
94. 'mal das cadeiras'; 'mal de caderas'.
95. amastigote.
96. *Wohlfahrtia magnifica*; *Oestrus ovis*; *Chrysomya bezziana*; *Cuterebra* sp.
97. chlorine
98. *Plasmodium falciparum*; *P. vivax*; *P. ovale*; *P. malariae*; *P. knowlesi*; *P. falciparum.*
99. (erythrocytic); liver.
100. *Legionella; Mycobacteria; Listeria; Francisella.*

4 True or False Questions

Read each statement given below and indicate in the corresponding brackets whether it is [true] or [false].

1.	Cats shed *Toxoplasma gondii* oocysts multiple times in their lifetime.	[..........]
2.	Schistosome eggs can be detected only in faeces.	[..........]
3.	In pregnant cats, *Toxocara cati* larvae migrate across the placenta to infect the fetuses.	[..........]
4.	When using ivermectin for the treatment of demodicosis, do not give to dogs under 12 weeks old, do not administer with amitraz, and continue treatment for 1 month beyond two negative skin scrapings taken 2 weeks apart because treatment may take months to achieve a clinical remission.	[..........]
5.	Leukopenia and thrombocytopenia, side effects associated with treatment with pyrimethamine, can be mitigated via the administration of leucovorin.	[..........]
6.	Benznidazole is the drug most commonly used to treat Chagas disease.	[..........]
7.	Human ingestion of *Taenia saginata* eggs can lead to cysticercosis.	[..........]
8.	Neutrophils are the most important cells that regulate the outcome of *Leishmania* infection.	[..........]
9.	*Psoroptes ovis* mites are characterized by having, on their first and second pair of legs, a three-segmented pedicle that ends in funnel-shaped suckers.	[..........]
10.	The juvenile *Dicrocoelium dendriticum* flukes migrate up directly from the intestine to the common bile duct.	[..........]

11.	Dogs with p-glycoprotein defect are hypersensitive to ivermectin and may develop neurotoxicosis and even death due to accumulation of ivermectin in their brain.	[.........]
12.	Feline demodicosis is caused by infestation with *Demodex cati* and *D. gatoi*.	[.........]
13.	Mites differ from insects by being wingless and by having four pairs of legs (adult and nymphal stages), instead of the three pairs found in insects.	[.........]
14.	Although both *Pulex irritans* (human flea) and *Xenopsylla cheopis* (rat flea) lack pronotal and genal combs, the human flea can be discriminated from the rat flea by having undivided mesopleuron.	[.........]
15.	Finding *Cysticercus fasciolaris* in mice means that these mice must have ingested dog faeces.	[.........]
16.	Black water fever is a manifestation of malaria caused mainly by *Plasmodium vivax*.	[.........]
17.	*Dictyocaulus viviparus* is the common lungworm found in horses.	[.........]
18.	Demodicosis in dogs is caused by *Demodex canis*, and can be transmitted from dams to their newborn pups.	[.........]
19.	Male circumcision can reduce transmission of *Trichomonas vaginalis* from male to female partners.	[.........]
20.	Visceral and ocular larva migrans are associated with infection with *Trichinella spiralis*.	[.........]
21.	Hypobiosis and overwintering are similar concepts in nematode biology and are underpinned by the same mechanism.	[.........]
22.	The first human case infected with *Angiostrongylus cantonensis* was reported in 1945 from Taiwan.	[.........]
23.	Fipronil should not be used in rabbits, as adverse reactions and even death could occur.	[.........]
24.	Canine hepatozoonosis is a protozoal disease transmitted to dogs via the bite of the brown dog tick (*Rhipicephalus sanguineus*).	[.........]
25.	The 'alveolar' cysts of *Echinococcus multilocularis* have more invasive ability, but contain less fluid than those of *E. granulosus*.	[.........]
26.	Skin scrapings collected from a mangy animal can be enriched for mites by dissolving in 10% potassium hydroxide (KOH) solution.	[.........]
27.	The FAMACHA system is used for assessment of anaemia.	[.........]
28.	There is a connection between flea infestation and tapeworm infection.	[.........]
29.	Transmission of *Eimeria* oocysts among chickens occurs via direct faecal-oral transmission from fresh droppings.	[.........]
30.	The acaricide amitraz should not be used on horses.	[.........]

31.	Removal of a tick is better performed by grasping the tick as close to the host skin as possible with fine-tipped forceps (or tweezers) and gently pulling upward.	[.........]
32.	When *Toxoplasma gondii* enters a host cell, it immediately starts to incorporate parasite DNA into the host cell's chromosomes.	[.........]
33.	'Gid or Sturdy' is a neurological disease in sheep caused by infection of the brain with the larva of *Taenia ovis*.	[.........]
34.	The obligate intracellular protozoan parasite *Toxoplasma gondii* lives inside a vesicle inside the surrogate host cell cytoplasm known as 'phagosome'.	[.........]
35.	*Schistosoma mansoni* egg has a terminal spine.	[.........]
36.	Ivermectin's toxicity in mammals is attributed to the similarity between the invertebrate glutamate-activated chloride channels and the vertebrate GABAA-gated chloride channels.	[.........]
37.	Soft ticks (Argasidae) lack a scutum and their mouthparts cannot be seen when the tick is viewed from above.	[.........]
38.	The ciliate *Balantidium coli* is the largest known protozoan that infects humans.	[.........]
39.	Skin wart-like lesions caused by female cat fleas (*Ctenocephalides felis*) often occur in resource-limited populations in Latin America, the Caribbean and sub-Saharan Africa.	[.........]
40.	The Congo floor maggots are the only know blowfly larvae to feed on human blood.	[.........]
41.	Direct faecal-oral transmission of *Cyclospora cayetanensis* can occur from relatively fresh faeces.	[.........]
42.	Once ticks are dead on the host, they fall off the host's body immediately to the ground.	[.........]
43.	Transtadial transmission of protozoa through an arthropod is the passage of a protozoan from the egg to the adult arthropod stage.	[.........]
44.	Dogs contribute to the transmission of *Toxoplasma gondii*.	[.........]
45.	*Felicola subrostratus* is a biting louse of cats and has a unique triangular-shaped head.	[.........]
46.	The poultry red mite, *Dermanyssus gallinae*, is the most damaging parasite of laying hens worldwide.	[.........]
47.	Otodectic mange (also known as otoacariasis) is caused by *Otodectes cynotis* mite, which infests the ear canals of cats and dogs.	[.........]
48.	The location of trypanosomes in the body of the insect vector does not influence the transmission of the infective stage.	[.........]
49.	House dust mite allergens can contribute to the development of allergic diseases.	[.........]
50.	Veterinarians should warn their clients about the potential hazards to humans from fleas.	[.........]

Answers: True or False Questions

1. **False**. Cats shed *Toxoplasma gondii* oocysts only once in their lifetime.

2. **False**. Schistosome eggs can be detected in faeces (e.g. *Schistosoma mansoni*) and urine (e.g. *Schistosoma haematobium*).

3. **False**. In pregnant queens, *Toxocara cati* larvae cannot cross the placenta and cannot infect fetuses in utero.

4. **True**.

5. **True**.

6. **True**.

7. **False**. It is the ingestion of *Taenia solium* eggs, which can result in a condition called 'cysticercosis'. The human is the accidental intermediate host and the cysticerci develop in a number of different human tissues, including the brain and muscles.

8. **False**. Macrophages and dendritic cells are the two main antigen-presenting cells (APCs) that mediate resistance and susceptibility of the host during *Leishmania* infection.

9. **True**.

10. **True**.

11. **True**.

12. **True**.

13. **True**.

14. **True**.

15. **False**. Finding *Cysticercus fasciolaris* in mice means that these mice must have ingested cat faeces.

16. **False**. Black water fever is a manifestation of malaria caused mainly by *Plasmodium falciparum*.

17. **False**. *Dictyocaulus arnfieldi* is the common lungworm found in horses.

18. **True**.

19. **True**.

20. **False**. Accidental infection with the larvae of *Toxocara canis* can cause visceral and ocular larva migrans.

21. **False**. Although both hypobiosis and overwintering are mechanisms used by some parasitic nematode species to survive during adverse climatic conditions, they represent different concepts. Hypobiosis describes the inhibition of larval development inside the animal, which is associated with reduced parasite metabolism. Overwintering occurs outside the body of the animal (e.g. on pasture) and is used by some larvae to survive through cold, dry weather, or any other adverse conditions that would make parasite survival difficult.

22. **True**.

23. **True**.

24. **False**. Dogs become infected by ingesting a brown dog tick containing oocysts. The infective sporozoites exit from oocysts and undergo schizogony in many tissues of the dog, including skeletal muscle.

25. **True.**
26. **True.**
27. **True.**
28. **True.** The tapeworm *Dipylidium caninum* uses fleas as an intermediate host. Cats or dogs (definitive hosts) are infected when they ingest infected fleas.
29. **True.**
30. **True.** Amitraz is not recommended for use on horses due to adverse effects.
31. **True.**
32. **False.** When *Toxoplasma gondii* enters a host cell, it immediately starts to form a membranous compartment called the 'parasitophorous vacuole', separating the proliferating parasites from the host-cell cytoplasm.
33. **False.** Coenurosis (also known as Gid or Sturdy) is a neurological disease that occurs in sheep when *Coenurus cerebralis* (the larval stage of *Taenia multiceps*) invade and colonize their brain.
34. **False.** *Toxoplasma gondii* lives inside the surrogate host cell cytoplasm within a parasitophorous vacuole.
35. **False.** *Schistosoma mansoni* egg has a lateral spine. *Schistosoma haematobium* eggs are passed in the urine and have a prominent terminal spine.
36. **True.**
37. **True.**
38. **True.**
39. **False.** Sand fleas, also known as 'jigger, chigger or chigoe, fleas, are haematophagous insects that can attack humans and domestic animals. They produce wart-like lesions in the skin of the affected host and often occur in resource-limited populations in Latin America, the Caribbean and sub-Saharan Africa.
40. **True.** *Auchmeromyia luteola* (Congo floor maggot) is distributed in tropical Africa. Their larvae are the only larvae known to feed on blood. The females lay the eggs on mud floors, usually contaminated by urine. The larva attacks sleeping humans and injures the intact skin until it reaches the blood vessels.
41. **False.** *Cyclospora cayetanensis* oocysts shed in faeces require a period of sporulation time (days to weeks) outside the host to become infective. Therefore, direct faecal-oral transmission from fresh faeces does not occur.
42. **False.** Dead ticks don't always fall off the host immediately because they release a cement-like substance that can take up to 5 days to dissolve even after the tick has died.
43. **True.**
44. **True.** Dogs are coprophagic animals and can act as mechanical carriers of *Toxoplasma gondii* oocysts.
45. **True.**
46. **True.**
47. **True.**

48. False. The location of trypanosomes in the body of the vector determines the mode of transmission of the infective stage. Salivarian transmission occurs via the insect proboscis, whereas in stercorarian transmission, the infective stage is deposited in faeces on the host's skin while the vector is feeding and enters the body of the host through a break in the skin or through mucous membranes.

49. True. The faecal pellets of several mite species are the main source of allergens in the house dust.

50. True. Because fleas can bite humans and some humans become hypersensitive. Also, cat scratch disease caused by *Bartonella henselae* can cause anything from local pustules to severe illness.

5 Image-Based Questions

(**I**) The given figure shows unstained preparation of the branchiuran *Argulus* sp. (Muller, 1785) isolated from skin of fresh water fish. Identify the structures labelled A to G.

Taxonomy:
Phylum: Arthropoda
Class: Crustacea
Subclass: Branchuira
Order: Arguloidea
Family: Arguloidae
Genus: Argulus

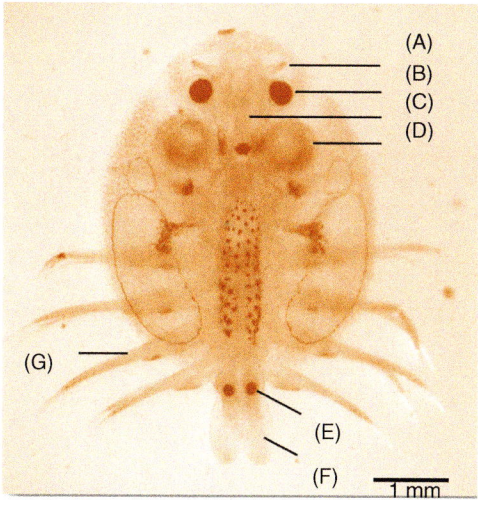

Fig. 5.1.

©CAB International 2019. *555 Questions in Veterinary and Tropical Parasitology* (H.M. Elsheikha and X-Q. Zhu)

(**II**) This figure illustrates the life cycle of *Leishmania infantum*. Identify each stage of the life cycle labelled A to E.

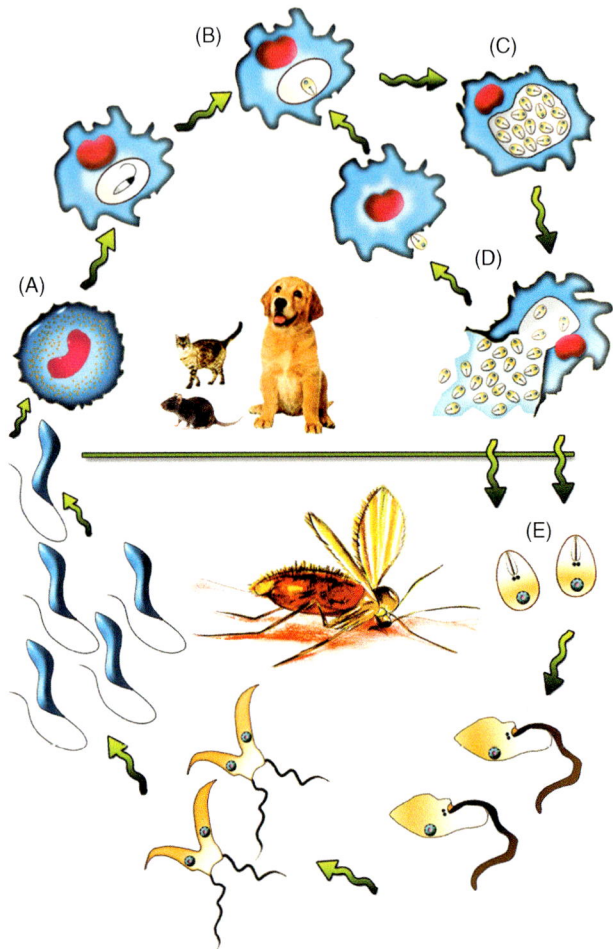

Fig. 5.2.

Fig. 5.2.

(III) These schematic drawings illustrate trophozoites and cysts of different *Entamoeba* spp. Identify the species name of each stage and briefly describe its characteristic morphometric features.

Table 5.1.

Characteristic

Trophozoites
Size
Number and
shape
of nuclei

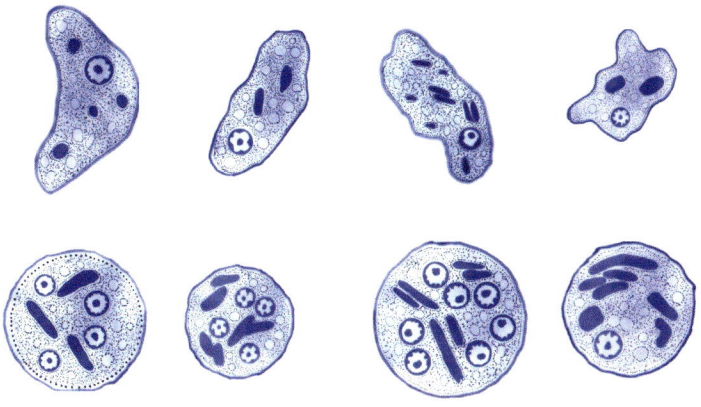

Cysts
Size
Number of
nuclei
Appearance

(IV) The given figure shows a collection of different stages of different parasite species. Identify each of the following parasite structures labelled A to L.

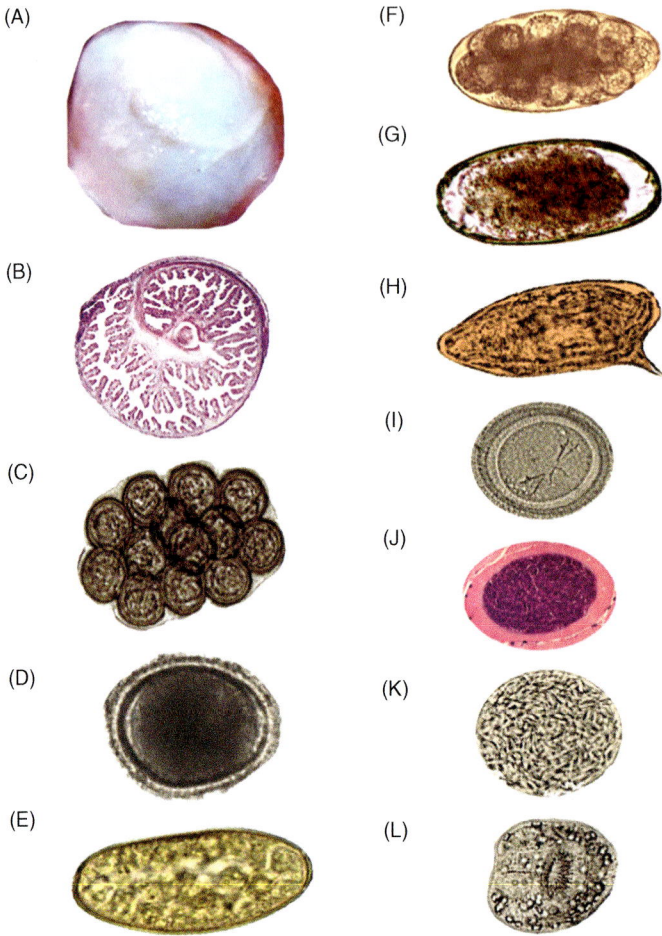

(A)

(B)

(C)

(D)

(E)

(F)

(G)

(H)

(I)

(J)

(K)

(L)

Fig. 5.3.

(V) This figure illustrates the life cycle of fleas. Identify each of the following stages of the life cycle labelled A to D.

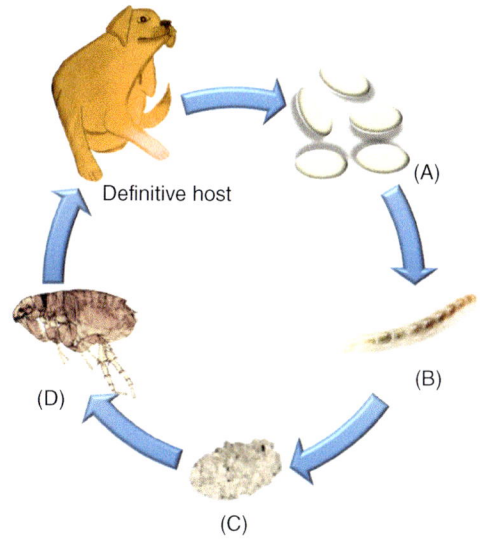

Fig. 5.4.

(VI) This figure illustrates the life cycle of liver flukes. Identify each of the following stages of the life cycle labelled A to E.

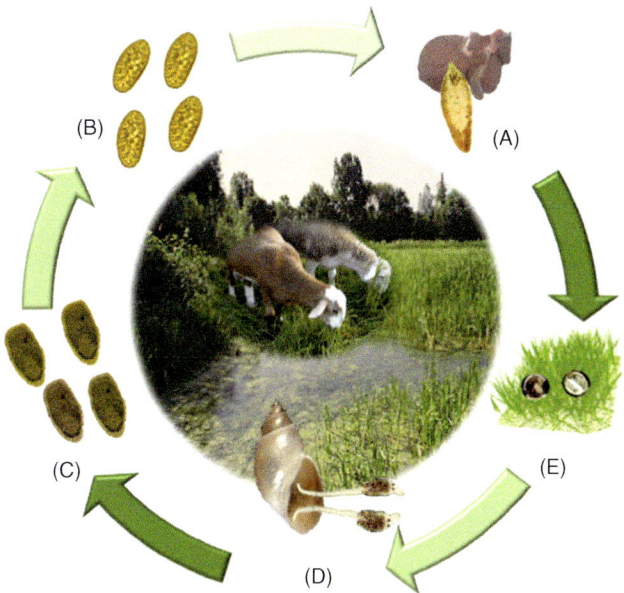

Fig. 5.5.

(**VII**) This figure shows the microscopic anatomical structure of a tapeworm. Identify each of the following organs labelled A to H.

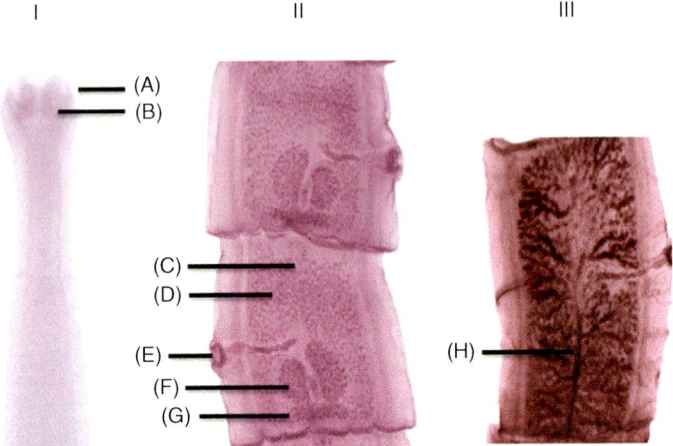

Fig. 5.6.

(**VIII**) The given photographs show an adult Ixodid tick – dorsal and ventral view. Identify each of the following parts labelled A to E.

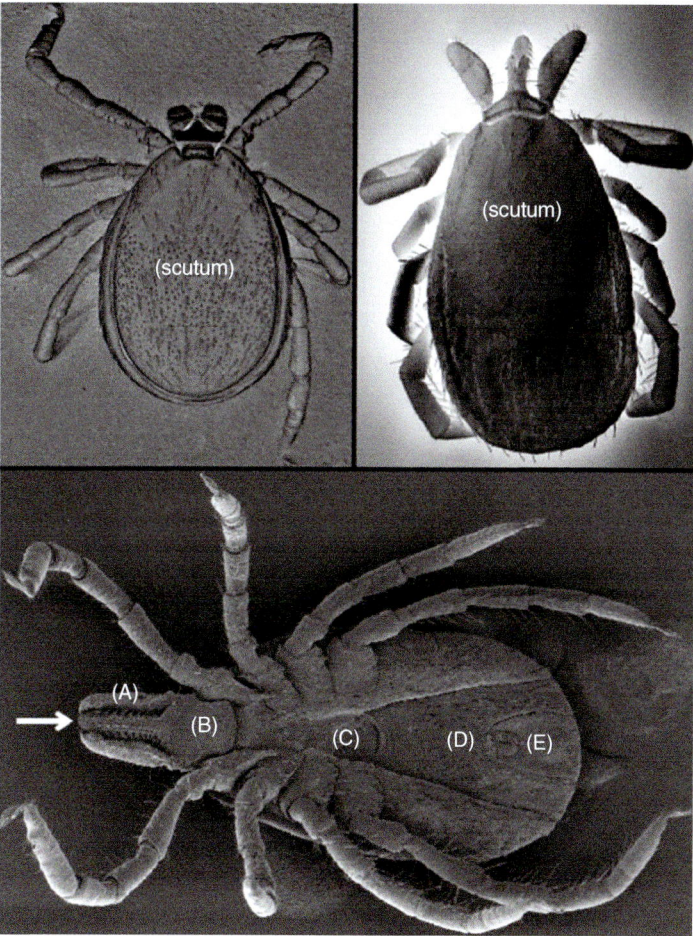

Fig. 5.7.

Answers: Image-based Questions

(**I**) Morphological features of *Argulus* sp. (Muller, 1785).

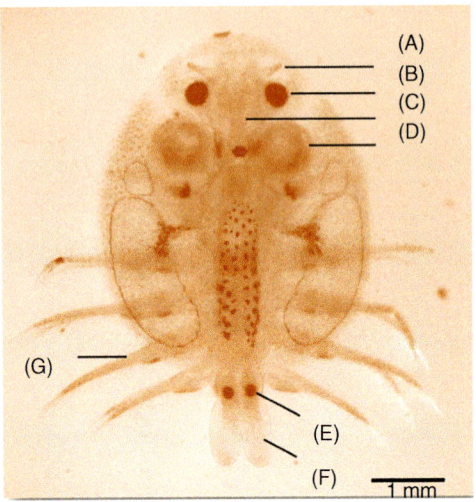

Remarks: Argulus species, also known as fish louse, is an ectoparasite found on the external body surface and fins of fish. The disease caused by *Argulus* infestation is called 'argulosis'. Female is larger than male. Key morphological features include: (A) antenna, (B) eye, (C) proboscis, (D) sucker, (E) testis, (F) abdomen, and (G) thoracic leg.

(II) Life cycle stages of *Leishmania infantum*.

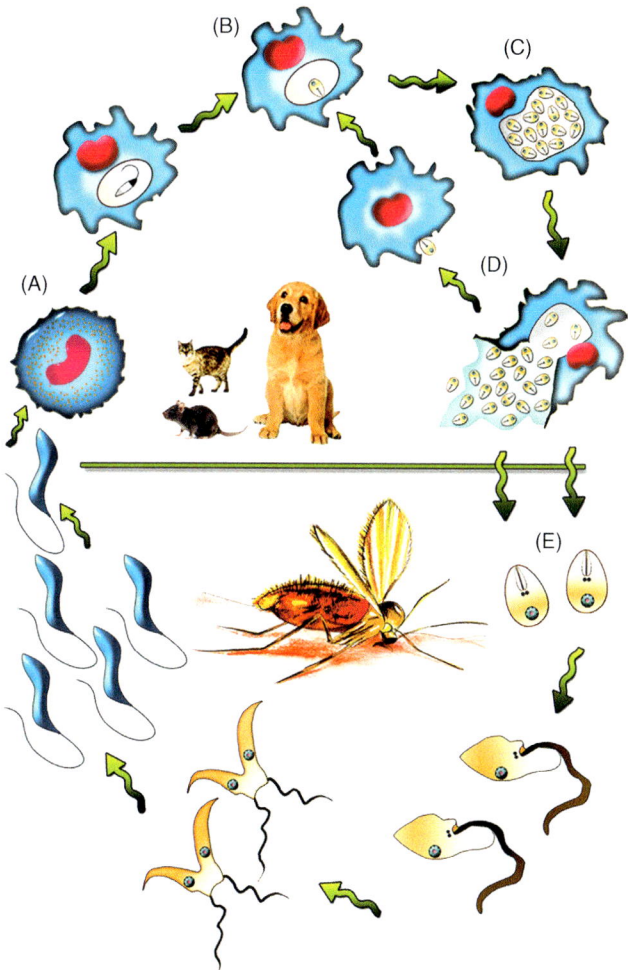

Fig. 5.2.

Remarks: Cats or dogs act as the definitive (final) host and become infected when they are bitten by a sandfly. (A) promastigotes are inoculated into the host's dermis and colonize the resident macrophages, where they develop in to amastigotes (B) and proliferate within phagolysosome (C). (D) after the rupture of macrophages the released amastigotes infect new macrophages. If the host fails to control the infection in the skin, amastigotes disseminate via lymphatics and the blood to infect the reticulo-endothelial system. (E) amastigotes that are ingested by a female sandfly during feeding transform back into promastigotes, which replicate within the sandfly to complete the life cycle.

103

(III) Morphometric characteristics of trophozoites and cysts of *Entamoeba* species.

Table 5.1.

Characteristic	E. histolytica/E. dispar/E. moshkovskii	E. hartmanni	E. coli	E. polecki
Trophozoites				
Size	15–60 µm	4–12 µm	15–50 µm	15–20 µm
Number and shape of nuclei	1*	1*	1**	1***
	Nucleus has evenly arranged chromatin on the nuclear membrane and centrally located karyosome	Nucleus resembles a "bull's eye" with even chromatin and central karyosome	Nucleus with coarse chromatin irregularly placed on nuclear membrane with eccentric large karyosome	Nucleus is similar to that in *E. histolytica*

Cysts				
Size	10–25 µm	5–10 µm	10–35 µm	10–15 µm
Number of nuclei	*mature cyst*: 4 *immature cyst*: 1 or 2	*mature cyst*: 4 *immature cyst*: 1 or 2	8 (occasionally 16)	Often 1, rarely 2 or 4
Appearance	Chromatoidal bars have smooth, rounded edges	Chromatoidal bars are similar to those in *E. histolytica*	Chromatoidal bars are splinter-shaped and irregular	Chromatoidal bars like those in *E. histolytica.* Inclusion body can be seen.

*difficult to see in unstained preparation; **often visible in unstained preparation; ***occasionally seen on wet preparation.

(**IV**) Identification of different stages of different parasite species.

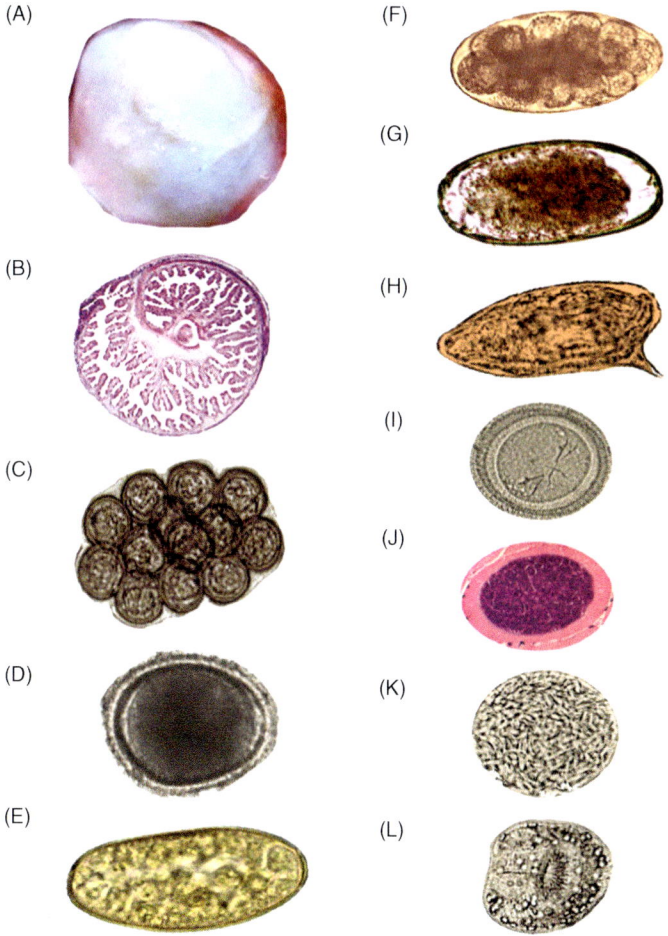

Fig. 5.3.

Remarks: (A) *Echinococcus granulosus* (hydatid) cyst (10 cm in diameter). (B) cysticercus/bladder worm (1 cm in diameter). (C) *Dipylidium caninum* egg capsule (30–50 µm x 25–48 µm). (D) *Toxocara canis* egg (80–85 µm x 70–80 µm). (E) *Fasciola* species egg (130–150 µm x 60–90 µm). (F) *Strongylus* species egg (80–90 µm x 40–50 µm). (G) *Ancylostoma caninum* egg (60–75 µm x 35–40 µm). (H) *Schistosoma mansoni* egg (120–150 µm x 50–75 µm). (I) *Taenia* species egg (30 x 40 µm in diameter). (J) *Sarcocystis* species cyst (50 µm in diameter). (K) *Toxoplasma gondii* cyst (50 µm in diameter). (L) hydatid sand, also known as protoscolex, collected from hydatid fluid within *Echinococcus* cyst (100 µm in diameter).

(V) Life cycle stages of fleas.

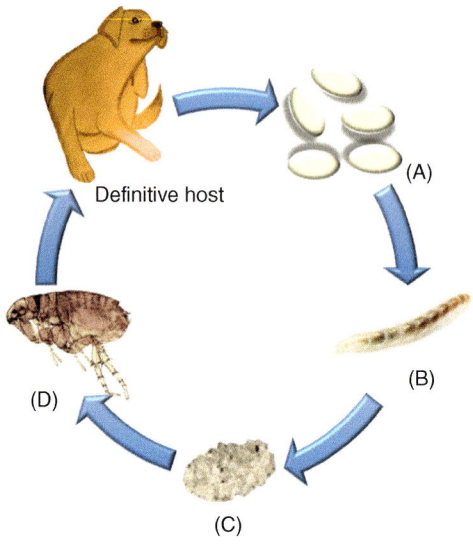

Fig. 5.4.

Remarks: (A) eggs, pearlescent white or translucent oval shape with rounded ends. They fall to the floor near nests and bedding. (B) eggs hatch into maggot-like larvae, which are the most active and sensitive stage. They feed on organic debris, exfoliated skin, flea eggs, but particularly faeces of the adult fleas, which is known as 'flea dirt' and contains undigested blood. (C) the 3rd instar larvae produce a whitish, loosely spun silk-like cocoon inside which they transform to pupa. (D) adult fleas emerge from within the cocoon if they are triggered by stimuli such as heat, physical pressure, and carbon dioxide.

(**VI**) Life cycle stages of liver flukes.

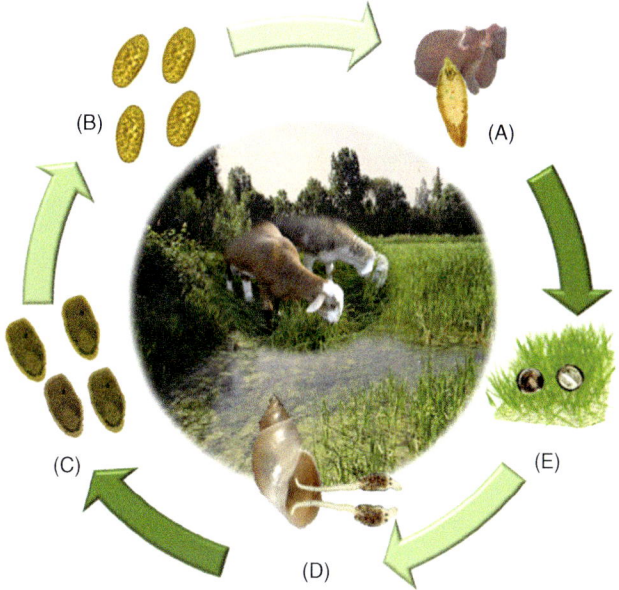

Fig. 5.5.

Remarks: Liver fluke (*Fasciola* spp.) infection has been estimated to contribute nearly $3 billion annually in lost productivity across sheep and cattle farms, worldwide. Animals feeding on contaminated pastures where the encysted metacercariae are found can become infected with *Fasciola* flukes, which advance inside the body of the animal until they reach and colonize the liver of infected animals, leading to the disease known as fasciolosis. This water-related parasitic disease cannot be efficiently controlled by using flukicide drugs alone. (A) adult flukes reach sexual maturity in the liver of the definitive vertebrate host. (B) eggs (130–150 μm x 60–90 μm) are shed in manure. (C) miracidia hatch from eggs and penetrate mud snail of *Lymnea* sp. (known as *Galba truncatula*), where sporocysts and rediae grow inside the snail tissue. (D) cercariae exit from the snail and encyst on grass as metacercariae (E).

107

(**VII**) Key anatomical features of tapeworms.

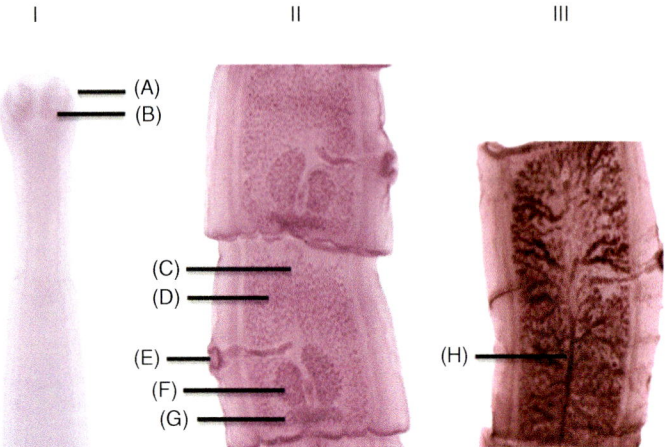

Fig. 5.6.

Remarks: Key anatomical features of a tapeworm include:

I. scolex (A), which contains attachment organs, such as suckers (B).
II. Mature segments, which contains uterus (C), testes (D), common genital pore (E), ovaries (F), vitelline gland (G).
III. Gravid segment, which contains gravid uterus (H) full of fertilized eggs.

(**VIII**) Key morphological features of adult Ixodid tick, dorsal and ventral view.

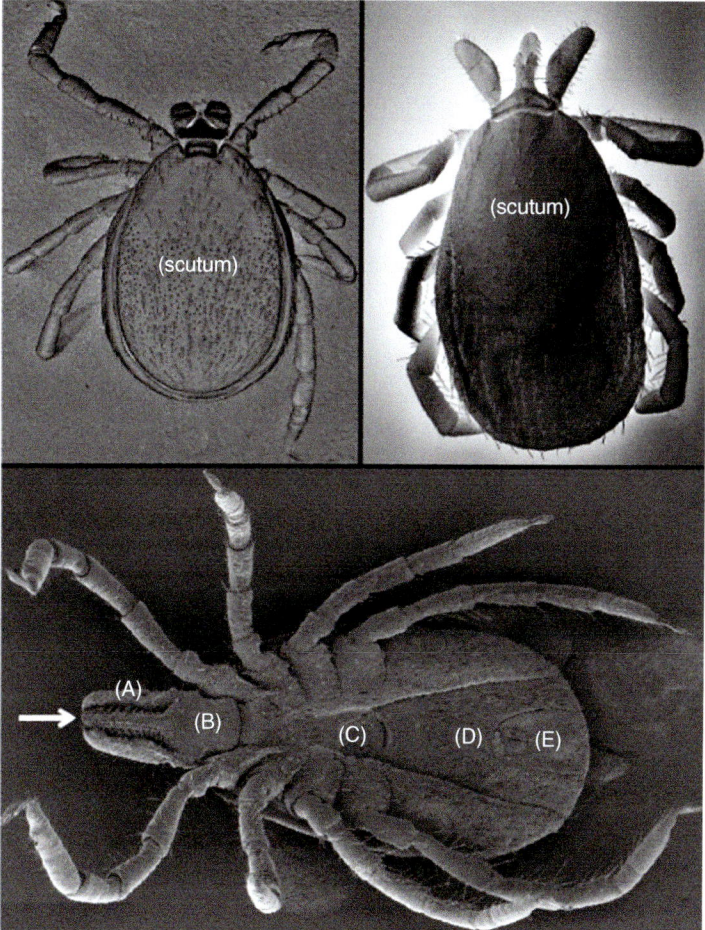

Fig. 5.7.

Remarks: Scutum (shaded) covers entirely the dorsum of adult male *Ixodes ricinus* (top left), but covers partially the dorsum of the female tick (top right). Key morphological features include: hypostome (arrow), pedipalps (A), capitulum (B), general aperture (C), and pre-anal groove (D) arching anteriorly to the anus (E). About 865 tick species exist worldwide, of which ~650 species belong to the family Ixodidae. The genus *Ixodes* involves ~245 species; of these 14 are in the ricinus complex.

CABI – who we are and what we do

This book is published by **CABI**, an international not-for-profit organisation that improves people's lives worldwide by providing information and applying scientific expertise to solve problems in agriculture and the environment.

CABI is also a global publisher producing key scientific publications, including world renowned databases, as well as compendia, books, ebooks and full text electronic resources. We publish content in a wide range of subject areas including: agriculture and crop science / animal and veterinary sciences / ecology and conservation / environmental science / horticulture and plant sciences / human health, food science and nutrition / international development / leisure and tourism.

The profits from CABI's publishing activities enable us to work with farming communities around the world, supporting them as they battle with poor soil, invasive species and pests and diseases, to improve their livelihoods and help provide food for an ever growing population.

CABI is an international intergovernmental organisation, and we gratefully acknowledge the core financial support from our member countries (and lead agencies) including:

 Ministry of Agriculture
People's Republic of China

 Australian Government
Australian Centre for
International Agricultural Research

 Agriculture and
Agri-Food Canada

 Ministry of Foreign Affairs of the
Netherlands

 Schweizerische Eidgenossenschaft
Confédération suisse
Confederazione Svizzera
Confederaziun svizra

Swiss Agency for Development
and Cooperation SDC

Discover more

To read more about CABI's work, please visit: **www.cabi.org**

Browse our books at: **www.cabi.org/bookshop**,
or explore our online products at: **www.cabi.org/publishing-products**

Interested in writing for CABI? Find our author guidelines here:
www.cabi.org/publishing-products/information-for-authors/